Kalam Mir

Arsenic in Plants: Methods of Extraction and Speciation

Kalam Mir

Arsenic in Plants: Methods of Extraction and Speciation

New Insights in The Challenges Faced with Arsenic in Terrestrial Plants

Copyright © Kalam Mir 2019

Cover: Kalam Mir

ISBN: 9781092932219

v

Abstract

A sequential arsenic extraction method was developed with approximately double the extraction efficiency (EE) of the current methods for terrestrial plants. In this method, plants were extracted first by 1:1 water-methanol followed by 0.1 M-HCl. Total arsenic in plant and soil samples were determined by ICP-AES and HG-AAS, and arsenic speciation was done by HPLC-HGAAS. Results showed overall suppression of extraction of inorganic arsenic but preference for As(V) by methanol. Dilute HCl was efficient in extracting arsenic from plants; however, determination of organic species was difficult in this media. Sequential extraction with 1:1 water-methanol followed by 0.1 M-HCl was most useful in extracting and speciating both organic and inorganic arsenic from plants. Although both organic and inorganic arsenic compounds could be detected simultaneously in synthetic gastric fluid extracts (GFE), it lacked the general efficiency required for extracting arsenic from various terrestrial plant samples.

A systematic study was conducted using the developed sequential arsenic extraction method and many plant species from an abandoned gold mining area at Deloro, Ontario, Canada. Total arsenic in plants ranged 1-500 $\mu g.g^{-1}$. All plants accumulated arsenic up to the middle of the growing season and a few continued accumulating until the end of the season. Plants without roots overall contained the highest amounts of arsenic in the leaves. Assimilation in terms of As(V)/As(III) ratios in plants was independent of total arsenic absorbed by the plants. EE appeared to be inversely related to As(III) concentration in plants. DMA was extracted more efficiently by 100% methanol. Different trends of EE for different plant species were observed.

Plant extracts were analyzed by HPLC-HGAAS and HPLC-ICPMS; the agreement of results showed that the use of HPLC-HGAAS would be sufficient for the speciation of arsenic in plants grown on arsenic contaminated soils. XANES analysis indicated seasonal variation in arsenic species in plants. Multi-element analysis showed correlation between concentrations of arsenic and sulfur and calcium, and a number of other heavy metals in plants grown on arsenic contaminated soils.

Claims to Originality

In the course of developing an efficient arsenic extraction method, the following work, results and observations were accomplished, which, to the best of the author's knowledge, were original and not reported before. This study reported or demonstrated for the first time:

1. The development of a sequential method of arsenic extraction from plants. The developed method extracted arsenic with water-methanol followed by dilute HCl and was highly efficient (approximately 100% more efficient than the traditional water-methanol method). The method was shown to preserve the integrity of the arsenic species and was simple to follow and apply for routine and large-scale applications (Chapter 3, Chapter 4, Appendix A).

2. Dilute (0.1 M or 0.05 M) HCl was a much better arsenic extractor than the traditional water-methanol; and 100% and 50% methanol were more suitable for the organoarsenic species in the presence of much higher inorganic arsenic content in plants. Three novel schemes for arsenic extraction were proposed in this work (Section 3.3.2, Fig. 3.3, Section 3.15, Fig. 3.12, Section 5.2, Fig. 5.4, and Table 5.2).

3. The first large scale investigation of arsenic in plants grown on arsenic contaminated soils at Deloro in Ontario, Canada. This work reported the first systematic study of a large number of Deloro plants over three seasons (Chapter 4).

4. The extraction efficiency of arsenic from plants was inversely related to the As(III) concentration in plants (Section 4.8.1, Fig. 4.3).

5. The extraction efficiency of arsenic from plants was species dependent and showed the various trends of arsenic extraction from plants (Section 4.9, Fig. 4.5).

6. Water-methanol preferentially extracted As(V) over As(III) and methanol resisted extraction of inorganic arsenic from plant matrices. This was a reason for the low extraction efficiency of water-methanol solvent (Section 4.8.1, Table 4.13; Section 5.2.1, Fig. 5.4).

7. The seasonal variation of arsenic species in plants by XANES (Section 5.4.3, Fig. 5.9). The results of arsenic speciation in the original plant samples, residues of plants from extraction, and the plant extracts using XANES and HPLC-HGAAS methods were also reported in this study (Section 5.4.2, Fig. 5.8, Section 5.4.4, Table 5.5).

8. Observed correlation between concentrations of As and S, and As and Ca, in plants grown on arsenic contaminated soils (Section 5.5.1, Fig. 5.10). Correlation between As-Fe, As-Co and As-Cd in these plants was also found in this work (Section 5.5.2).

Acknowledgements

In my long journey through time and space for completing this Ph.D. thesis, I have been indebted to many in many ways. I would like to thank all who have supported and encouraged me. Without your support, it would not be possible for me to come to this point. I thank you all from the bottom of my heart.

I would like to thank Dr. Allison Rutter, Director of the Analytical Services Unit (ASU) and Adjunct Professor, School of Environmental Studies, Queen's University for her patience and dedication in guiding me through this arsenic research project.

I would like to thank Dr. John Poland, former Director of ASU, for taking me to the historic Deloro gold mining areas to collect plant and soil samples for arsenic analysis. I will remain grateful to John for accepting me as a graduate student in his wonderful laboratory.

I have my sincerest gratitude for Dr. Gary vanLoon, Professor of Chemistry, for his support and guidance to me throughout this journey in the Department of Chemistry at Queen's University. I also thank Dr. David Wardlaw, Professor and Chairman of Chemistry, for his support. Financial support, in the form of a research assistantship, from the Department of Chemistry, the School of Environmental Studies, and the School of Graduate Studies and Research is deeply appreciated.

An important part of this project was performed at the laboratory of the Environmental Sciences Group (ESG) at the Royal Military College of Canada at Kingston for which I am thankful to Dr. Kenneth Reimer, Director of ESG.

I am greatly indebted to Dr. Iris Koch at ESG for her guidance and support throughout this research project. A number of key experiments of this thesis would not have been possible without her active help and support. Dr. Koch's extensive work on arsenic has been a guideline of this research on arsenic.

The technical support that I have received from the staff at the ASU and the ESG was tremendous. A key factor to the success of this Ph.D. project was their untiring help and support in performing my experiments in these laboratories. I thank Dr. Graham Cairns at the ASU for keeping the ICP-AES working for running my samples. Mary Andrews never said no to any of my requests and always knew where to find what I was looking for. Paula Whitley, the ASU Laboratory Manager, always managed to find room for me to perform my experiments even in the busiest schedules of the ASU work. The only thing they do not know is how very thankful I am for their help and support at the ASU through these years.

I am deeply thankful to Paula Smith, Ph.D. student, Angela and Leigh at the ESG for their help and support in running my experiments at the ESG. Paula took my samples to the Argonne National Laboratory in Illinois and showed me how to use the WinXAS program for XANES experiments. I thank Dr. Bubby Kettlewell for painstakingly reading this thesis and making corrections and suggestions for improvement in the meaning and readability of the texts. There are so many I do not name but I am grateful for their support and company throughout this long journey in completing this Ph.D. thesis. I thank you all very much for your support, encouragement and kind words.

Finally, I am thankful to my wife, Yamun Nahar, for her support and patience to bear with me in my 'arduous' journey toward a doctorate. My ten-year-old daughter, Progga Adiba Mir, was always persistent in reminding me of my studies when I would indulge in a little too much TV or music. I will be forever grateful to her for this.

List of Tables

Table 1.1. Arsenic compounds commonly detected in the environment

Table 2.1. Instrument parameters for arsenic determination

Table 2.2. Determination of arsenic in certified reference materials: seaweed, bush branches and leaves, and citrus leaves by nitric acid digestion and ICP-AES detection

Table 3.1. Arsenic in soil samples collected from arsenic waste tailing and abandoned goldmine areas at Deloro

Table 3.2. Total arsenic determined by ICP-AES and HG-AAS. Results are mean ± 1 s.d. from 3 independent determinations

Table 3.3. Total arsenic determined by wet ashing by HNO_3 (HNO_3 Method) and dry ashing in oven and digesting by 1:3 v/v HNO_3/HCl (ASU method)

Table 3.4. Mass balance experiment performed with water-Soxhlet extracts and residues of plants. Results are mean ± 1 s.d. from 3 independent determinations

Table 3.5. Mass balance experiment for Yellowknife plants. Arsenic concentrations in the extracts and residues were determined by ICP-AES. Concentrations are in $\mu g.g^{-1}$

Table 3.6. Arsenic in unfiltered acid digested and filtered undigested extracts of P-6. Results are mean ± 1 s.d. from 3 independent determinations

Table 3.7. Total arsenic in low arsenic containing plant samples determined by HG-AAS. Results are average ± 1 s.d. from 3 independent determinations

Table 3.8. Results of spike recovery experiments: inorganic and organic spikes were determined by HG-AAS and ICP-AES, respectively. Results are average ± 1 s.d. from 3 independent determinations

Table 3.9. Speciation of arsenic extracted from plant samples. MA was detected only in the water-methanol extracts of P-1 (1.51 ± 0.04). Results are average ± 1 s.d. from 3 independent determinations, concentrations are in $\mu g.g^{-1}$

Table 3.10. Reliable detection limits (RDL) of arsenic species determined by HPLC-HGAAS. RDL = 2(t-stat)(s.d.)

Table 3.11. Determination of method detection limits (MDL) of ICP-AES and HG-AAS. MDL = (t-stat)(s.d.)

Table 3.12. Determination of the effect of soaking on extractability of arsenic. All concentrations are in $\mu g.g^{-1}$. Percent extraction efficiencies are given in parentheses.

Table 3.13. Comparison of arsenic extraction and speciation by 0.3 M H_3PO_4 with those of sequential and 1.0 M-HCl. Concentrations are in $\mu g.g^{-1}$

Table 3.14. Comparison of extraction efficiencies of the sequential, 1.0 M-HCl, 0.3 M-H_3PO_4 and 0.3 M-H_3PO_4 (de-watered) extractions determined by ICP-AES and HPLC-HGAAS

Table 3.15. Results of 100% methanol, 50% methanol and water extracts by HPLC-HGAAS. Results are mean ± ave. deviation from 2 independent determinations. All concentrations are in $\mu g.g^{-1}$ and void spaces represent species not detected

Table 3.16. Total arsenic extracted by the solvents. Results are mean ± ave. deviation from 2 independent determinations. Concentrations are $\mu g.g^{-1}$

Table 3.17. Comparison of extraction efficiencies of 100% MeOH and 50% MeOH for DMA with statistical data

Table 3.18. Arsenic extracted by GFE and 0.05 M-HCl. pH of 1.8 was used for both extraction media. Results are mean ± 1 s.d. from 3 independent determinations

Table 3.19. Effect of HCl concentrations on extraction efficiency of gastric fluid

Table 3.20. Effect of pepsin concentration with and without salt on extraction efficiency of gastric fluid. The test plant was field horsetail with a total As content 122 $\mu g.g^{-1}$. Results are mean ± ave. deviation from 2 independent determinations

Table 3.21. Pepsin-HCl-salt (PHS) extraction of plant samples. EE's of PHS and sequential extractions are reported for comparison. Results are mean ± ave. deviation, n = 2 for plants DL-207 and DL-209

Table 3.22. Sequential schemes and 1:1 methanol-0.1 M-HCl extraction results. A and A' are sequential extraction of Scheme A; B and B' of Scheme B; and C of Scheme C from Figure 3.15. nd = not detected

Table 3.23. Step by step extraction of As from plant P-6 (smooth horsetail, 24 $\mu g.g^{-1}$ total As) by 0.05 M-HCl-sonication process. Sample was 0.5 g (± 0.1 mg). Results are mean ± 1 s.d. from 3 independent determinations

Table 3.24. Effect of sample weight on extraction efficiency (EE). Time of sonication for each step is indicated in the table. Results are mean ± 1 s.d. from 3 independent determinations

Table 3.25. Effects of ratio of extractant volume to sample mass, extractant volume, and size of extraction tube on extraction efficiency (EE). One-step 30 minutes and one-step 20 minutes (as indicated) of sonication were employed. Results are mean ± 1 s.d. from 3 independent determinations

Table 3.26. Extraction of arsenic from Yellowknife plants by sequential extraction method. Extraction efficiencies of 1:1 water-methanol (traditional method of plant extraction) and sequential method are compared. Results mean ± ave. deviation from two independent analyses for a number of plants are reported for quality assurance

Table 3.27. Speciation of arsenic in the plant extracts by HPLC-HGAAS (see Table 3.26). Total As in the extracts was determined by ICP-AES. Mean ± ave. deviation from 2 independent determinations are reported for plants YK-6 anf YK-11. Concentrations are in $\mu g.g^{-1}$

Table 4.1. Total Arsenic in Deloro plants determined by ICPAES after acid digestion is given. A number of plants were analyzed in duplicate and results are listed as (mean ± ave. deviation). Plants collected in May were given one/two-digit ID Nos.; in July and September were given three digit, 100 and 200 ID Nos., respectively

Table 4.2. Soil arsenic concentrations and mean, median and ranges of plant arsenic concentrations in samples collected from six sites of Deloro gold mine area

Table 4.3. Determination of the variation of arsenic concentrations in plants and soils collected from one square meter plots of Deloro gold mining area

Table 4.4. Total arsenic in various plant species collected from different sites of Deloro, ON, Canada

Table 4.5. Seasonal variation of arsenic concentrations in plants collected at different times of the season. Arsenic in same plant species collected from same site are compared. For visual inspection of the changes see Figure 4.1

Table 4.6. Determination of arsenic in different parts of plants. For plant scientific names refer to Table 4.1

Table 4.7. Sequential extraction and speciation of plants collected in May 2003 from Deloro. No DMA in these plants but 0.10 $\mu g.g^{-1}$ MA was detected in DL-60

Table 4.8. Sequential extraction and speciation of plants collected in July 2003 from Deloro. Organoarsenic species were detected in water-alcohol extracts only. 0.10 $\mu g.g^{-1}$ MA was detected only in DL-124. Mean ± ave. deviation from two independent analyses of a number of plants are reported for QA

Table 4.9. Sequential extraction and speciation of plants collected in September 2003 from Deloro. Organoarsenic species were detected in alcohol extracts only

Table 4.10. Additional plants from Deloro sampled in May, July and September 2003 were extracted and speciated for organoarsenic species. Mixed spike recovery experiment was carried out along with these plants and results are reported in this table and discussed in Chapter 3.7.1

Table 4.11. The speciation results of As in 1:1 water-methanol and 0.1 M-HCl extracts of plants from Deloro are reported separately. EE's of two methods are compared. No DMA or MA was detected in HCl medium. Mean ± ave. deviation from duplicate determinations are provided to show QA.

Scientific and common English names of the plants are given in Table 4.1. Void spaces indicate no species detected

Table 4.12. Comparison of the overall extraction efficiency of the 1:1 water-methanol traditional method and sequential method

Table 4.13. Ratios of arsenate to arsenite in 1:1 water-methanol, 0.1 M-HCl and overall inorganic arsenic in the plants calculated from data reported in Table 4.11

Table 5.1. Extraction efficiencies of 1:1 water-MeOH (1:1), sequential and 0.1 M-HCl single solvent (SS) extractions. %EE of SS extractions after adding the organoarsenic species are also reported in the table for comparison

Table 5.2. Three extraction schemes of arsenic from plants. Extraction results of 100% methanol and 0.1 M-HCl, 50% methanol and 0.1 M-HCl, and 0.1 M-HCl as single solvent (SS) extractants are compared. Arsenic concentrations are in $\mu g.g^{-1}$

Table 5.3. Comparison of HPLC-HGAAS (HGAA) and HPLC-ICPMS (ICP) results of arsenic species in 100% methanol and 50% methanol extracts of Deloro plants. <LOD indicates concentration of species was close but not above detection limit. All concentrations are in $\mu g.g^{-1}$

Table 5.4. XANES analysis of Deloro and Yellowknife plants. Results from XANES fit with As(III) and As(V), and As(III), As(V) and As(III)-Glu together are reported. ND indicates values insignificant or negative

Table 5.5. The results of As(III) and As(V) from XANES (fractions in solid ground sample) and HPLC-HGAAS (fractions in liquid extracts) analysis are reported.

Table 5.6. Results of arsenic, macro and minor elements in Deloro plants are reported. All Concentrations are in $\mu g.g^{-1}$

Table 5.7. Arsenic and other essential, beneficial and nonessential elements in plants are listed. Concentrations are in $\mu g.g^{-1}$

List of Figures

Figure 1.1. Chemical structures of common inorganic and organoarsenic species, their names and abbreviations

Figure 1.2. Systematic procedures for the determination of arsenic in terrestrial plant samples are given. (AAS = atomic absorption spectrometry (flame or flameless); HG = hydride generation; CT = cryogenic trapping; HPLC = high performance liquid chromatography; ICP = inductively coupled plasma; AES = atomic emission spectrometry; MS = mass spectrometry; IEC = ion exchange chromatography; ET = electrothermal; NAA = neutron activation analysis; DCP = direct current plasma; AFS = atomic fluorescence spectrometry; MIP = microwave induced plasma; CE = capillary electrophoresis; EQ = electrochemical methods; and UV-vis. = spectrophotometry). Adapted from references 3 and 58

Figure 3.1. Flow chart of arsenic extraction and total arsenic determination

Figure 3.2. Extraction efficiencies (EE) of different solvents in solvent-Soxhlet method for three plants are presented. Error bars are ± 1 s.d. from three independent determinations

Figure 3.3. Extraction efficiencies (EE) of different solvents in solvent-sonication method for three plants are shown. Error bars represent ± 1 s.d. from three independent determinations

Figure 3.4. Results of exhaustive extraction of arsenic from plant (P-6) by water-Soxhlet process are presented. Errors represent ± 1 s.d. from 3 independent determinations

Figure 3.5. Extraction efficiency as a function of solvent pH is shown. Error bars represent ± 1 s.d. from 3 independent determinations

Figure 3.6. Extraction efficiency as a function of total arsenic in plant matrices by Soxhlet-solvent process is shown. Error bars represent ± 1 s.d. from three independent determinations

Figure 3.7. Extraction efficiency as a function of total arsenic in plant matrices by solvent sonication process. Error bars represent ± 1 s.d. from 3 independent determinations

Figure 3.8. DMA calibration curves prepared with different arsenic calibration standards are presented. Error bars represent ± 1 s.d. from 3 independent determinations

Figure 3.9. Representative HPLC-HGAAS chromatograms. A: Standard arsenic species; B: DMA spike; C: MA spike; and D: 1:1 water-methanol extract of plant P-1 (field horsetail). HPLC was performed with anion exchange column (Hamilton PRP-X100 250 x 4.6 mm column), 20 mM ammonium phosphate, pH 6.0 at 1.5 mL.min^{-1}

Figure 3.10. Ratio of arsenate to arsenite extracted in dilute HCl solutions as a function of total arsenic in the plants

Figure 3.11. Determination of loss of water from plant matrices as function of time of drying in open air and heating in oven at 105°C. Fifteen Yellowknife plant samples were dried for this experiment. The plants are listed in detail in Table 3.26

Figure 3.12. HPLC-HGAAS chromatograms of 1:1 water-methanol (A), water (B) and 0.1 M-HCl (C) extracts of field horsetail (P-1). HPLC of water extracts (B), acid extracts (C) was performed in June (same day). HPLC of 1:1 water-methanol extracts (A) was performed in August of the same year. HPLC conditions are given in Figure 3.9

Figure 3.13. The bar graphs show amounts of arsenic extracted as function of polarity of the solvents

Figure 3.14. HPLC-HGAAS chromatograms (Abs. Vs. Time in Sec.) of Yellowknife and Deloro plants extracted in 100% methanol, 50% methanol and DDW. Nearly identical chromatograms (not shown here) were obtained for duplicate and triplicate independent analyses of the samples indicating stability of the arsenic species throughout the analytical procedure. HPLC conditions are described in Figure 3.9

Figure 3.15. Extraction schemes used for the optimization of the arsenic extraction methods are shown

Figure 4.1. Bar graphs show seasonal changes in arsenic concentrations in a number of plant species. Sampling sites are shown in parentheses. See Table 4.5 for details

Figure 4.2. The ratios of arsenate to arsenite are plotted against the total arsenic in plants and the total arsenic extracted by sequential method. Values of R^2 from linear regressions are shown for each category. A and B: field horsetails; C and D: panicle aster; and E and F: purple loosestrife

Figure 4.3. The ratios of arsenite to arsenate plotted against the arsenic extraction efficiencies of twelve plant species by the sequential method

Figure 4.4. Representative HPLC-HGAAS chromatograms of Deloro plants extracted in 1:1 water-methanol and 0.1 M-HCl sequentially. HPLC was performed with anion exchange column (Hamilton PRP-X100 250 x 4.6 mm column), 20 mM ammonium phosphate, pH 6.0 at 1.5 mL.min^{-1}

Figure 4.5. Extraction efficiency (EE) as function of total arsenic and plant matrix

Figure 5.1. Comparison of EE by1:1 water-methanol-0.1 M-HCl sequential and 0.1 M-HCl single solvent extractions of Yellowknife plants

Figure 5.2. Comparison of EE by1:1 water-methanol-0.1 M-HCl sequential and 0.1 M-HCl single solvent extractions of Deloro plants

Figure 5.3. Extraction efficiencies of Canada goldenrod and Purple loosestrife with respect to 0.1 M-HCl and sequential extractions

Figure 5.4. Extraction efficiencies of 50% methanol (1:1 v/v water-methanol) and 100% methanol

Figure 5.5. HPLC-ICPMS anion exchange chromatograms of: (A) standard (20 µg.L^{-1}) solutions of arsenic compounds, (B) Deloro plant DL-207 and (C) DL-119. ICPMS confirmed MA in DL-207 and DMA in DL-119 determined by HPLC-HGAAS. Other organoarsenic species were not detected in these plants by either method

Figure 5.6. XANES spectra (Photon energy (keV) vs. Normalized absorption (a. u.)) of standard As(III) and As(V) species

Figure 5.7. XANES spectra (Photon Energy (keV) vs. Normalized Abs. (a. u.)) of dry ground plant samples and corresponding residues from sequential extractions are presented. A: Dry ground sensitive fern, B: Residue of sensitive fern; C: Dry ground field horsetail, D: Residue of field horsetail; and E: Dry ground purple loosestrife, F: Residue of purple loosestrife

Figure 5.8. As(III) and As(V) in fresh frozen plant and dry ground sample by XANES analysis. DL-201: sensitive fern (Onoclea sensibilis); DL-207: field horsetail (Equisetum arvense) and DL-220: purple loosestrife (Lythrum salicaria)

Figure 5.9. Comparison of fractions of As(III) and As(V) in original dry ground plant samples and residues from sequential extractions. Extraction efficiency of each plant species is given in parentheses

Figure 5.10. Variations in the species of arsenic in a number of plant species during a growing season

Figure 5.11. Plots of arsenic concentrations vs. calcium and sulfur concentrations in plants from Deloro

Figure 5.12. Cluster plot of arsenic and iron in plant samples

Table 5.13. Cluster plot of arsenic and cobalt in plant samples

Figure 5.14. Cluster plot of arsenic and cadmium in plant samples

List of Map and Photos

Map 4.1. Map shows Deloro gold mining area (Courtesy Ministry of Environment (MOE), ON, Canada). Plant and soil sampling sites are shown in boxes

Photo 3.1. A sample of field horsetail (*Equisetum arvense*)

Photo 3.2. A sample of smooth horsetail (*Equisetum laevigatum*)

Photo 3.3. The remnants and reminders of 100 years of gold mining activities at Deloro

Photo 3.4. Arsenic tailing pond near sampling sites 4, 5 and 6

Photo 4.1. An aerial view of the historical Deloro gold mining area is shown. (Courtesy Ministry of Environment (MOE), ON, Canada)

Photo 4.2. A view of leached water from the 'covered' arsenic tailings flowing down the stream to form shallow pond shown in Photo 4.3

Photo 4.3. A view of shallow water body formed by leached water and rain water run-off over the arsenic tailings around the gold mining sites

Photo 4.4. A view of dead vegetation strewn by the stream running with arsenic contaminated water

Photo 4.5. An area near plant sampling Sites 1, 2, and 3 north of the Young's Creek and south of the arsenic tailing area is shown

Photo 4.6. A general view of the arsenic sampling Sites 4, 5 and 6 near the industrial area and Deloro village is shown

Photo 4.7. A view of the washed plants lay on the laboratory table for drying

Photo 4.8. A photo of the author of the thesis 'facing arsenic without fear' is shown near the netted arsenic tailing pond in the background

List of Abbreviations*

AAS	Atomic Absorption Spectroscopy
AB	Arsenobetaine
AC	Arsenocholine
AES	Atomic Emission Spectroscopy
As	Arsenic
As(III)	Arsenite
As(V)	Arsenate
CRM	Certified Reference Material
DDW	Distilled Deionized Water
DL	Deloro
DMA	Dimethylarsinic acid
EE	Extraction Efficiency
EXAFS	Extended X-ray Absorption Fine Structure
GFE	Gastric Fluid Extraction
Glu	Glutathione
HG	Hydride Generation
HPLC	High Pressure (Performance) Liquid Chromatography
ICP	Inductively Couple Plasma
MA	Methylarsonic acid
MeOH	Methanol
MS	Mass Spectrometry
PC	Phytochelatin
PHS	Pepsin Hydrochloric acid and Salt extraction

TETRA	Tetramethylarsonium ion
TMAO	Trimethylarsine oxide
XANES	X-ray Absorption Near Edge Spectroscopy
YK	Yellowknife

*Note: some of these acronyms are used in combination, e.g. HGA-AS, ICP-AES.

xx

Arsenic in Plants: Extraction and Speciation

Contents

CHAPTER 1: INTRODUCTION 4
1.1. ARSENIC 4
1.2. ARSENIC FROM NATURAL AND ANTHROPOGENIC SOURCES 4
   1.2.1. Arsenic Compounds in the Environment 5
1.3. THE TOXICITY OF ARSENIC: A GLOBAL PROBLEM 6
1.4. ARSENIC SPECIES IN THE TERRESTRIAL ENVIRONMENT 10
   1.4.1. Arsenite (As(III)) and Arsenate (As(V)) 10
   1.4.2. Abandoned Goldmines at Deloro, Ontario, Canada 11
1.5. ANALYTICAL METHODS 12
   1.5.1. Methods of Arsenic Detection and Quantitation 12
1.6. METHODS OF ARSENIC SPECIATION 16
   1.7.1. Arsenic Species in Terrestrial Plants 18
   1.7.2. Extraction Methods for Arsenic in Plants 19
      1.7.2.1. Solvent Extraction 19
      1.7.2.2. Enzyme and Gastric Fluid Extractions 21
1.8. THE RESEARCH OBJECTIVES OF PRESENT STUDY 22
CHAPTER 2: METHODS 24
2.1. SAMPLING AND ANALYSIS 24
   2.1.1. Sample Collection and Preparation 24
   2.1.2. Plants from Yellowknife 25
   2.1.3. Preparation of Soil Samples 25
   2.1.4. Digestion of Plant Samples 26
   2.1.5. Digestion of Soil Samples 26
2.2. ANALYTICAL TECHNIQUES USED 27
   2.2.1. Total Arsenic by ICP-AES 27
   2.2.2. Total Arsenic by HGAASS 28
   2.2.3. Arsenic Speciation by HPLC-HGAAS 30
   2.2.4. Arsenic Speciation by HPLC-ICPMS 31
   2.2.5. X-Ray Absorption Near Edge Structure (XANES) of Arsenic Species - Fingerprints from XANES 31
2.3. QUALITY ASSURANCE AND QUALITY CONTROL (QA/QC) 32
CHAPTER 3: ARSENIC EXTRACTION AND SPECIATION IN PLANTS GROWN ON ARSENIC POLLUTED SOILS - DEVELOPMENT OF SEQUENTIAL EXTRACTION METHOD 34
3.1. INTRODUCTION 34
3.2. TOTAL ARSENIC IN SOIL AND PLANT SAMPLES 35

3.2.1. Total Arsenic by Dry-Ashing and Wet-Ashing Methods 36
3.3. EXTRACTION OF ARSENIC FROM PLANTS: SOXHLET AND SONICATION PROCESSES
.... 40
   3.3.1. Solvent-Soxhlet Extraction Method... 40
   3.3.2. Solvent-Sonication Extraction Method.. 41
   3.3.3. Soxhlet and Sonication Processes: Results and Discussion............................... 43
3.4. THE EFFECT OF ARSENIC CONCENTRATION ON EXTRACTION EFFICIENCY (EE) 46
3.5. MASS BALANCE OF EXTRACTED AND UNEXTRACTED SAMPLES........................ 48
3.6. EFFECT OF FILTRATION ON ARSENIC CONCENTRATION.. 50
3.7. SPIKE RECOVERY AND SPECIATION 51
   3.7.1. Recovery of Mixed Spikes... 53
3.8. DMA EXPERIMENT 54
3.9. SPECIATION OF ARSENIC IN PLANT EXTRACTS 56
3.10. LIMITS OF DETECTION AND PRECONCENTRATION OF ARSENIC SPECIES 60
3.11. THE WATER CONTENT OF PLANTS... 61
3.12. ADSORBED OR ABSORBED ARSENIC IN PLANT MATRIX 63
3.13. EFFECT OF SOAKING ON EXTRACTION EFFICIENCY (EE)..................................... 64
3.14. EXTRACTION WITH ORTHOPHOSPHORIC ACID 65
3.15. EXTRACTION BY 100% METHANOL, 50% METHANOL AND WATER 67
3.16. GASTRIC FLUID EXTRACTION (GFE)... 73
3.17. EXTRACTION BY A COMBINATION OF ACID AND ENZYME 74
   3.17.1. The Enzyme in Various HCl Concentrations 75
   3.17.2. Effect of Pepsin Concentration and Salt on Extraction 76
   3.17.3. Extraction of Plants by Pepsin-HCl-Salt Solution (PHS) 77
3.18. SEQUENTIAL ARSENIC EXTRACTION METHOD: DEVELOPMENT 79
   3.18.1. Optimization of HCl-Sonication Extraction 79
      3.18.1.1. Schemes of Extraction: Flowchart 79
      3.18.1.2. Effect of Number of Extraction Steps, Sample Mass, Volume/Sample Mass Ratio, Size and Shape of Extraction Vessel and Sonication Time on EE 82
   3.18.2. The Development and Optimization of Sequential Method: Conclusions 84
3.19. SEQUENTIAL ARSENIC EXTRACTION METHOD: APPLICATION TO REAL SAMPLES 87
CHAPTER 4: ARSENIC EXTRACTION AND SPECIATION IN PLANTS FROM DELORO, ONTARIO, CANADA 93
4.1. INTRODUCTION 93
4.2. THE DELORO SITE 93
4.3. SAMPLE COLLECTION AND PREPARATION 97
4.4. TOTAL ARSENIC IN PLANTS AND SOILS 99
   4.4.1. Sampling Variations of Arsenic Concentration in Plants and Soils 108
4.5. ARSENIC IN DIFFERENT PLANT SPECIES 111
4.6. VARIATION IN ARSENIC CONCENTRATION IN PLANTS DURING THE GROWING SEASON 113
4.7. ANALYSIS OF ARSENIC IN DIFFERENT PARTS OF PLANTS 115
4.8. EXTRACTION AND SPECIATION OF ARSENIC BY SEQUENTIAL METHOD 116
.... 117
   4.8.1. Extraction of As(III) and As(V) by Sequential Method: Their Influence on the Extraction Efficiencies 125
   4.8.2. Organoarsenic Compounds in Deloro Plants 128
4.9. EXTRACTION OF ARSENIC: DEPENDENCE ON PLANT MATRIX 129
.... 132

4.10. ARSENIC IN THE EQUISETUM (HORSETAIL) SPECIES 132
CHAPTER 5: EXTRACTION OF ARSENIC - FURTHER ASPECTS 134
5.1. INTRODUCTION 134
5.2. THE WATER-METHANOL, SEQUENTIAL AND SINGLE SOLVENT ARSENIC
EXTRACTION 134
   5.2.1. 100% Methanol Extraction of Arsenic in Plants 137
5.3. THE ORGANOARSENIC COMPOUNDS IN PLANTS: COMPARISON OF HPLC-HGAAS
AND HPLC-ICPMS RESULTS 140
5.4. XANES ANALYSIS OF DELORO PLANTS 142
   5.4.1. Arsenic Species in Fresh and Dried Ground Plants 146
   5.4.2. Arsenic Species in Dry Ground Plant samples and Corresponding Residues of Extraction:
Implication for Extraction Efficiency (EE) 148
   5.4.3. Variation of Arsenic Species in Season 149
   5.4.4. XANES and HPLC-HGAAS Speciation of Arsenic 151
5.5. ARSENIC AND OTHER ELEMENTS IN PLANTS GROWN ON ARSENIC IMPACTED SOILS
.... 152
   5.5.1. Correlation of Arsenic with Sulfur and Calcium in Plants 156
   5.5.2. Correlation between Arsenic and other Heavy Metals in Plants 158
CHAPTER 6: CONCLUSIONS AND FUTURE WORK 161
6.1. CONCLUSIONS 161
   6.1.1. Development of a Sequential Extraction Method for Arsenic in Plants 161
   6.1.2. Arsenic in Plants from Deloro by Sequential Method 163
   6.1.3. Further Aspects of Arsenic Extraction Study 165
6.2. PROPOSALS FOR FUTURE WORK 167
   6.2.1. Effect of Plant Matrix on Arsenic Speciation 167
   6.2.2. Extraction Efficiency and As(III) in Plants 167
   6.2.3. Field Horsetail as Arsenic Accumulator 168
REFERENCES 170

CHAPTER 1: INTRODUCTION

1.1. ARSENIC

The word 'arsenic' invokes fear in many,[1] and fittingly it seems, for it is food for many bacteria living in nature's shadows. Arsenic, termed as 'the king of poisons' or even better, 'the poison of kings', has perhaps swayed human history more than any other element. The development and demise of cultures and human generations have been linked to the use of arsenic.[2] The beneficial achievements of arsenic that have been known for centuries, on the contrary, are less savored. Many dreaded diseases like the Black Death, chronic myelocytic leukemia, boils, and venereal diseases, to name a few, have been treated with arsenic formulations[3]. However, more than 80% of all arsenic used by man was utilized in an environmentally dissipative manner and was included in substances like herbicides, chemical warfare agents, and drugs. Moreover, a large percentage of arsenic found in mined and smelted ores was released directly to the environment.[2]

1.2. ARSENIC FROM NATURAL AND ANTHROPOGENIC SOURCES

Chemically, arsenic (As) is a metalloid element[4] and it is also classed as one of the so-called heavy metals.[5] Due to its chemical bond strength to carbon and sulfur, arsenic has a broad chemistry involving these elements. The most prevalent oxidation states of arsenic in nature are As(III) and As(V) and the standard redox potential ($E°$) of the pair As(III)/As(V) is a moderate +0.57 V.[6,8] In spite of the modern day acumen of analytical methods, it seems that the debate regarding whether arsenic is just a poison, an environmental nuisance or another essential element will continue for some time.[3] Nature is the source of all arsenic and it is found everywhere on Earth. The quantification of the global element cycle for arsenic has been based on extensive research.[7] Arsenic is a general constituent of sulfide deposits; 2-3% by weight in copper and lead ores and as much as 11% by weight in gold ores.[8,9] It is ubiquitous in the environment and is the 20th most abundant element with an average concentration 2-3 mg.kg^{-1} in crustal rock.[6,10,11] Arsenic is a component of more than 245 minerals.[12]

Arsenic in Plants: Extraction and Speciation

The concentrations and forms of arsenic that are of concern to humans and the environment may occur as a consequence of both natural as well as anthropogenic activities.[6,13,14] Examples of the naturally elevated arsenic to alarming levels are common in places such as Meager Creek hot springs in British Columbia and Yellowknife in North West Territories (NWT) of Canada.[15,16] In recent times, arsenic from underground aquifers has caused serious health concerns in Bangladesh.[17,18] The combination of natural and human activities such as gold mining, river transportation of toxic wastes from industries and agricultural runoff has greatly contributed to the environmental hazard of arsenic.[19,20]

The source analysis of the toxic contaminants has received much attention because, without knowing the source, mitigation efforts may be futile. Studies to distinguish between natural and anthropogenic sources of metals including arsenic in the Irish Sea have shown that the geochemistry of sediments impacted by human activity was different from that of the unimpacted sediments of a similar natural background, thus enabling a distinction between the natural and anthropogenic sources of contamination.[21] Another study successfully used Al and V as the concentration normalizing elements in regression analysis to distinguish between natural and anthropogenic sources of heavy metals including arsenic in a lake drainage basin.[22] A similar study was carried out in an arsenic impacted area of Bangladesh to determine whether natural or anthropogenic sources of arsenic or both, were responsible for the contamination.[23]

1.2.1. Arsenic Compounds in the Environment

A large number of arsenic compounds have been identified in the environment and in biological organisms.[6,24] Since arsenic compounds occur widely and in high concentrations in marine systems, this area initially received the most attention. It was later found that most of the arsenic compounds in marine organisms are also present in terrestrial systems.[11] Inorganic arsenate (As(V)) and arsenite (As(III)) are the most prevalent followed by organic methylarsonate (MA) and dimethylarsinate (DMA) species in the terrestrial environment.[16,25] Other arsenic compounds commonly found in relatively small concentrations in the terrestrial environment are arsenobetaine (AB), arsenocholine (AC), trimethylarsine oxide (TMAO), tetramethylarsonium ion (TETRA) and the arsenic-consisting ribosides also called arsenosugars.[11,26] In the marine environment, however, the organoarsenic species such as AB, TETRA and AC are predominant. The organoarsenic species MA and DMA are usually found in the more stable As(V) oxidation form but the reduced forms of the species, MA(III) and DMA(III), have been identified.[27] A list of the common terrestrial arsenic compounds and their chemical structures are given in Table 1.1 and Figure 1.1, respectively.

1.3. THE TOXICITY OF ARSENIC: A GLOBAL PROBLEM

Arsenic has been known for hundreds of years and at one time in history was only considered a poison.[2] But not all forms of arsenic are toxic, and toxicity is dependent on dose, bioavailability, time of exposure and cellular metabolism of the toxicant.[28,29,30,31] Laboratory experiments with primary cultured rat astroglia has shown that the dose-related cellular toxicity or DNA damage, and toxic effect of arsenite is greater than that of arsenate. However, MA(V) and DMA(V) have been shown to have no toxic effect at micromolar concentrations on the same rat cell culture.[28] The examination of the pathophysiological and molecular effects of inorganic As(III) and As(V), and organic MA(V) and DMA(V) arsenicals, on mouse liver, a known target of arsenic, has shown dose-dependent toxic effects for all arsenic species tested. The toxic effects of the organic species were dissimilar to those of the inorganic species.[29]

Table 1.1. Biologically important arsenic compounds commonly detected in the environment[a].

Name of compound	Abbreviations	Chemical formula	Product of hydride generation
Arsenous acid, Arsenite	As(III)	$As(OH)_3$	Arsine, AsH_3; B.P. -55°C
Arsenic acid, Arsenate	As(V)	$AsO(OH)_3$	Arsine, AsH_3; B.P. -55°C
Methylarsonic acid	MA	$CH_3AsO(OH)_2$	Methylarsine, CH_3AsH_2; B.P. 2°C
Dimethylarsinic acid	DMA	$(CH_3)_2AsO(OH)$	Dimethylarsine, $(CH_3)_2AsH$; B.P. 36°C
Trimethylarsine oxide	TMAO	$(CH_3)_3AsO$	Trimethylarsine, $(CH_3)_3As$; B.P. 52°C
Tetramethylarsonium ion	TETRA	$(CH_3)_4As^+$	--
Arsenobetaine	AB	$(CH_3)_3As^+CH_2COO^-$	--
Arsenocholine	AC	$(CH_3)_3As^+CH_2CH_2OH$	--
Arsenic-containing ribosides	Arsenosugars	Various sugar structures	Note[b]

[a] Adapted from references 24, 26, 27 and 58.
[b] Production of arsines from arsenosugars has been reported.[77]

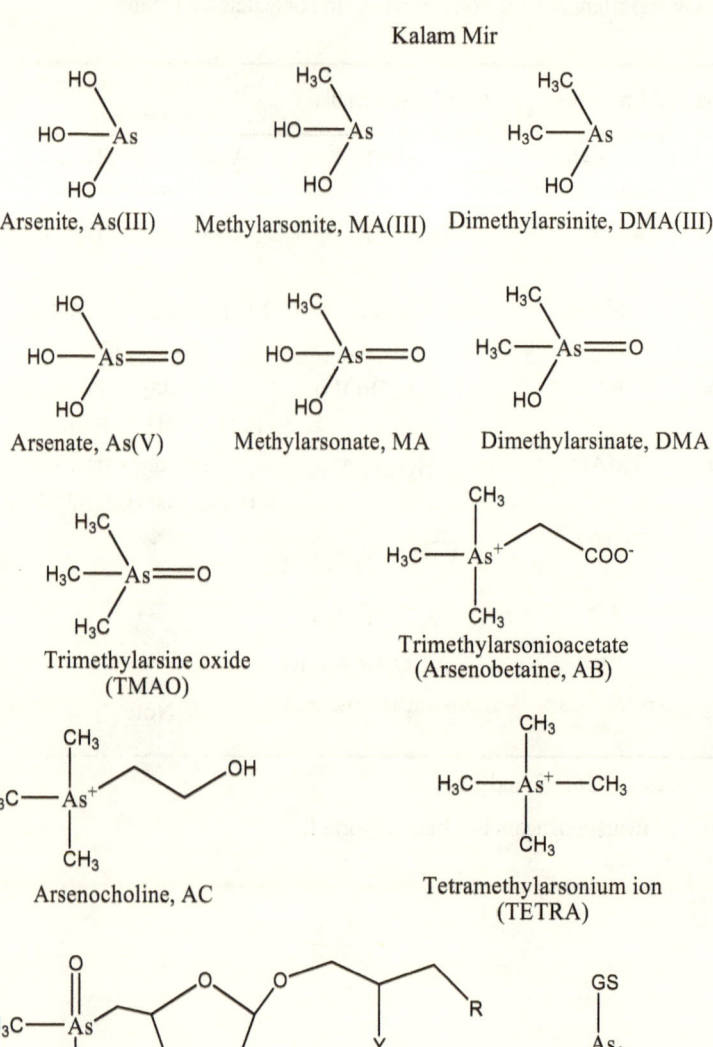

Figure 1.1. Chemical structures of common inorganic and organoarsenic species, their names and abbreviations.

The species-dependent higher toxicity of arsenic has been observed for all As(III) species with both inorganic and organic arsenicals. Furthermore, cytotoxicity tests using Chang human hepatocytes has shown that MA(III) was the most toxic of all arsenic species and revealed the order of toxicity as: MA(III) > As(III) > As(V) > MA(V) = DMA(V) for the Chang human hepatocytes.[30,32] The observation, though, has shed doubt on the earlier belief that methylation of inorganic arsenic was a detoxification process.[6] On the other hand, study with algal communities from lakes has indicated As(V), and to some extent MA(V), to be more toxic and, therefore, an environmental hazard.[33]

Hydroxyl radical mediated damage of DNA in keratinocytes by As(III) and adverse effects of arsenic exposure to the nervous system have been reported.[34,35] The absorption of arsenic has been found to occur mainly through ingestion from the small intestine with nominal absorption through the skin and by inhalation. Absorbed arsenic has been found to affect many organs such as the liver, kidney, skin and bladder, and is a well-documented carcinogen.[36,37,38]

The toxicity of arsenic to humans has been demonstrated on a massive scale. Millions of Bangladeshis and people in West Bengal, India have been unknowingly consuming arsenic-contaminated groundwater for decades.[17,39] Skin lesions in epidemic proportions,[40] respiratory effects,[41] hypertension,[42] cancer[43] and diabetes mellitus[44] have been shown to be associated with arsenic toxicity in the region. The correlation of diabetes mellitus and arsenic toxicity has been a recent observation and appears to be a global phenomenon since similar results have been found for people exposed to arsenic in Bangladesh, Sweden and Taiwan.[45] A relationship between severities of arsenic toxicity and nutritional deficiency has been observed in comparisons between populations in developing and developed countries; a significantly larger number of people from developing countries such as Bangladesh, China, India and Mongolia suffer from arsenic related neuropathy, skin lesions, vascular damage, and carcinogenesis.[46]

Arsenic poisoning through diet particularly of plant origin has been a global concern. Food crops from mining areas in Yellowknife, NWT in Canada have been found to be unsafe for human consumption.[25] An important observation has been the arsenic concentrations in the vegetables produced in Yellowknife residential gardens which are an order of magnitude higher than those grown in uncontaminated areas.[47] The concentration of arsenic in rice grown on paddy fields irrigated with arsenic-contaminated groundwater in Bangladesh has been found to be high and unsafe by a recent study.[48] Different food items collected from the villagers of an arsenic-impacted area in West Bengal, India and from different regions of Bangladesh have shown alarming levels of toxic arsenic that pose risks to large numbers of the population.[49,50]

The arsenic toxicity to plants or phytotoxicity depends significantly on the chemical form and concentration in the soil. The chemical similarity of arsenate to phosphate allows arsenic to become incorporated into many metabolic reactions and, As(III) by binding itself to the sulphydryl groups, interferes with plant biochemical and physiological processes.[51]

Arsenic toxicity has been shown to interfere with the carbohydrate metabolism in growing rice seedlings.[52] This is a significant finding in view of the fact that the toxicity might have potential adverse effects on the rice yield. It has been recognized in the early studies that only the soluble but not the total arsenic can be considered as an index of arsenic toxicity to plants and concentrations as low as 1 $\mu g.g^{-1}$ soluble arsenic may retard root and shoot growth of the arsenic sensitive species.[53,54]

1.4. ARSENIC SPECIES IN THE TERRESTRIAL ENVIRONMENT

1.4.1. Arsenite (As(III)) and Arsenate (As(V))

The most common forms of arsenic that exist in the environment are the oxyanions of arsenic; arsenite and arsenate (see Fig. 1.1). Sometimes referred to collectively as 'inorganic arsenic' these represent major forms (>90%) of arsenic in freshwater and terrestrial plants, but minor forms (<5%) in marine biota.[11] Many environmental agents, both abiotic and biotic, interconvert the two inorganic species and, for that reason, they are usually found together. Under normal environmental oxygen levels As(V) is the favored species.

The stability of the As(V)/As(III) redox pair in the environment is dependent on many natural processes. Consequently, integration of the physical, chemical and biological processes is necessary in order to predict the fate of As in real systems.[55] It has been shown that without external influence, such as bacterial and/or chemical degradation, the As(V)/As(III) pair is comparatively stable. An electrochemical reversibility study designed to test the As(V)/As(III) redox pair in aqueous conditions, typical of the environment, showed no chemical equilibrium between the As(V)/As(III) pair, unlike the well-known Fe(III)/Fe(II) redox pair, indicating the irreversible nature of the arsenic redox pair.[56] The nonequilibrium behavior of the As(V)/As(III) redox pair has been cited in several other studies.[55]

Many environmental factors, biological and/or non-biological, must be taken into consideration to explain As cycling in the environment.[6,55] The rates of individual reactions and processes like surface complexation, dissimilatory reduction, as well as the effects of extractant solvent, storage media, and speciation analysis must be taken into consideration to fully elucidate the arsenic reactions in plant extracts.

1.4.2. Abandoned Goldmines at Deloro, Ontario, Canada

Gold was discovered in 1866 in a southern Ontario community that later became known as Deloro. In fact, the word 'Deloro' in Spanish means 'Place of Gold'. The Deloro area is rich in mineral deposits and is situated where the Canadian Shield intersects the Great Lakes lowlands in Ontario, some 200 kilometers southwest of Ottawa and 65 kilometers east of Peterborough. Within a few years of the gold discovery, many shafts had been sunk and refining facilities had been constructed. There were about twenty gold mines by the mid 1890's operating at Deloro.[57] Although the gold mines in Deloro were closed in the early 1900s, the milling site was used to process silver, cobalt and other ores brought in from northern Ontario and other parts of the world during 1930s, 1940s and 1950s. Arsenic pesticides were also produced at the facilities from the arsenic byproducts of gold refining operations and continued until 1950s when organic pesticides were introduced.

A century of gold mining, and silver and cobalt ore processing has left severe and widespread environmental degradation in the Deloro area. The Moira River flows through the gold mining site and over time, arsenic-tailings were dumped close to the river. Contamination of the river water and ecosystem occurred by the runoff water leaching arsenic from the tailings. While government-initiated cleanup of hazardous ferric arsenate sludge is in progress, steps to determine the impact of the toxic wastes on the environment of the mining area have also been taken.[57] After taking control of the site in 1979, the Ministry of Environment (MOE), Ontario has taken specific steps to clean up the contaminated site. The MOE has established an arsenic treatment plant for groundwater that removes almost 99.5% arsenic and it has also established an onsite laboratory for wide-ranging surface and ground water monitoring. The MOE has also initiated off-site projects that include the 'Deloro Environmental Health Risk Study' and the 'Moira River Study' to assess the potential impact to the people and environment around the Deloro arsenic-polluted area. However, the toxicity to plant and wildlife of the area has not been evaluated.

1.5. ANALYTICAL METHODS

The analytical methods for arsenic have come a long way since James Marsh discovered his famous test for trace arsenic in cases of poisoning. With advanced methodologies and techniques,[3,6,58] the determination of total arsenic in many samples and different media has become routine. Historically, arsenic analysis began with the identification and/or determination of total arsenic in samples.[3] E. G. Mahin of Purdue University published in 1907 on the determination of total arsenic in London Purple by an AOAC method involving iodometric titration.[59] Even today, the knowledge of total arsenic is essential prior to doing any quantitative speciation analysis.

1.5.1. Methods of Arsenic Detection and Quantitation

The low-level determination of arsenic in modern times began with the advent of spectrophotometric methods such as the molybdenum blue method and the silver diethyldithiocarbamate method. However, the first method was selective for arsenate (As(V)) and the second one was selective for arsenite (As(III)). Electrochemical methods are capable of determining arsenic at trace levels but suffer from severe matrix interferences and lack robustness. These methods are not commonly employed for environmental arsenic determinations.[3]

Arsenic in Plants: Extraction and Speciation

The X-ray absorption methods, EXAFS (extended X-ray absorption fine structure) and XANES (X-ray absorption near edge spectroscopy), have been receiving acceptance for arsenic speciation analysis.[8,27,60,61] Since the chemical stability of arsenic species becomes vulnerable due to extraction and speciation processes, methods capable of doing speciation directly on samples (*in situ* analysis) may provide information on transformations of species occurring between the original sample and the extract. The XANES and EXAFS methods have been useful for unraveling the oxidation states and electronic environment, respectively, of arsenic in environmental samples.[60,62] These much desired methods, however, suffer from high detection limits and interference.[8,27] In spite of these limitations, results from these methods have complemented analyses by the more traditional methods.[61]

The analysis of arsenic began with the determination of total arsenic in samples[3] and due to the omnipresence of arsenic in the environment, the diversity of matrices where total arsenic has been determined is large. Depending on the sample matrices, both dry and wet ashing methods were employed for converting the arsenic of a sample in inorganic form.[63,64] Routine analysis for total arsenic is usually done by HGAAS, ICPAES or ICPMS. Extraction of arsenic will be discussed below for various matrices.

Hyphenated techniques such as the HPLC-HGAAS, HPLC-ICPMS, HPLC-HG-ICPMS, to name a few, have been successfully employed to determine arsenic species.[3,8,27,76] Systematic procedures for the determination of arsenic species in terrestrial plant samples are given in Figure 1.2. Among the detectors, ICPMS has found

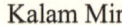

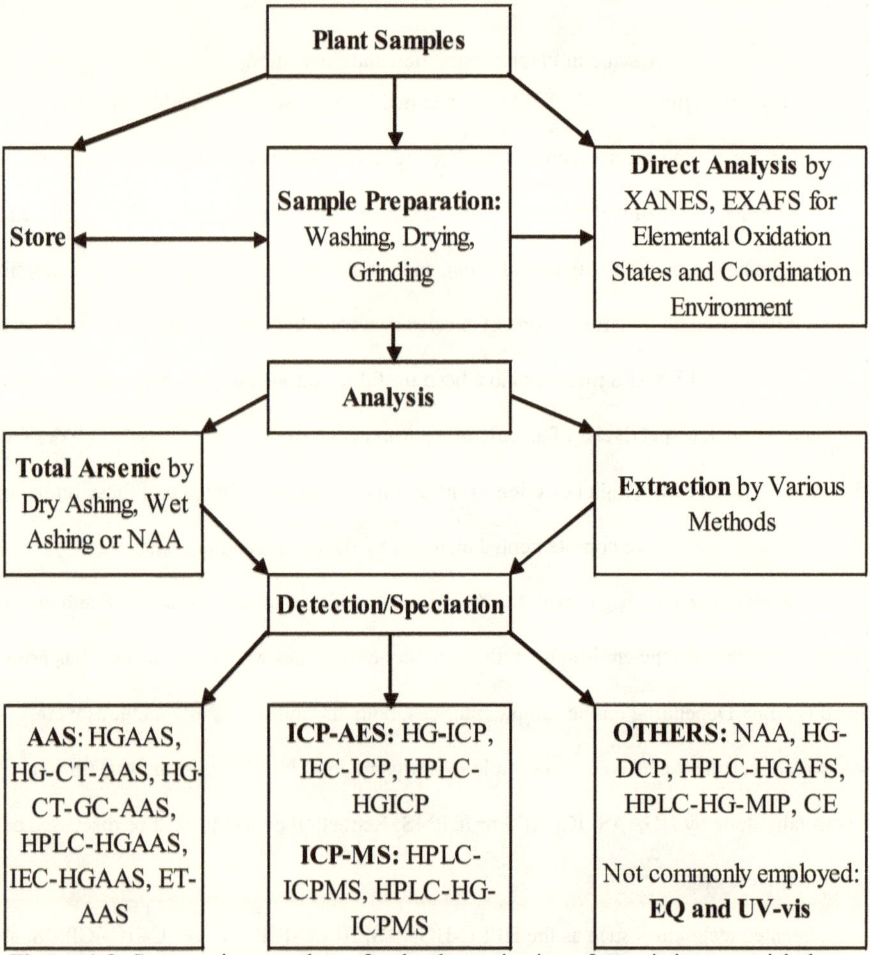

Figure 1.2. Systematic procedures for the determination of arsenic in terrestrial plant samples are given. (AAS = atomic absorption spectrometry (flame or flameless); HG = hydride generation; CT = cryogenic trapping; HPLC = high performance liquid chromatography; ICP = inductively coupled plasma; AES = atomic emission spectrometry; MS = mass spectrometry; IEC = ion exchange chromatography; ET = electrothermal; NAA = neutron activation analysis; DCP = direct current plasma; AFS = atomic fluorescence spectrometry; MIP = microwave induced plasma; CE = capillary electrophoresis; EQ = electrochemical methods; and UV-vis. = spectrophotometry). Adapted from references 3 and 58.

wide use in the identification of diverse arsenic compounds in terrestrial environment because of its excellent detection limits, wide linear range and multi-element capabilities.[3,26] For routine environmental monitoring and/or risk assessment many samples must be analyzed. For such real samples, only a few analytical methods such as the HPLC-HGAAS, HPLC-ICPAES and HPLC-ICPMS have been found to be suitable in terms of sensitivity, selectivity and ease of operation.[76] However, significant disadvantages of ICP based methods, particularly the ICPMS, are the high cost of purchase and expensive maintenance that preclude them being used by many laboratories.[3,76]

The other methods that have been used for the determination of total arsenic in soil samples are differential-pulse cathodic stripping voltametry[65,66] and nondispersive hydride-generation atomic fluorescence spectrometry.[67] Atomic absorption spectrometry with hydride generation and flow injection analysis coupled techniques were used to determine total arsenic in soil and plant matrices.[68] Sewage sludge was analyzed by alkaline oxidation and HGAAS.[69] Total arsenic in environmental-geological samples was determined by electrothermal atomic absorption spectrometry (ETAAS) using a tungsten furnace after solvent extraction and cobalt(III) oxide collection.[70] The total-reflection X-ray fluorescence (TXRF) spectrometry has been applied for analyzing elements including arsenic in water, soil, sewage sludge and biological samples. TXRF has the advantage of being calibrated without the need for standards and provides analytical results comparable to those from ICPAES methods.[71,72,73]

Comparative studies of a number of analytical techniques for the determination of arsenic have been reported.[74,75] In the analysis of peat for arsenic and antimony, the performances of instrumental neutron activation analysis (INAA), HGAAS, sector field ICPMS (SF-ICPMS) and quadruple-ICPMS (Q-ICPMS) have been compared. INAA has underestimated the values systematically, but results from HGAAS and SF-ICPMS have agreed very well. The best precision of results has been provided by the HGAAS and Q-ICPMS. However, ICPMS has used less amount of sample and determined more than one element of environmental importance.[74] In another study, comparison of HGAAS, ETAAS, ICPAES and ICPMS for the determination of arsenic in an environmental solid matrix has shown that, except for ETAAS with Zeeman background correction, each method has had its own peculiar drawbacks and has not offered any particular advantage over the other techniques.[75]

1.6. METHODS OF ARSENIC SPECIATION

Because of broad differences in the chemical properties of the arsenic compounds, such as anionic, cationic and neutral arsenic species existing within the same sample, various separation procedures are required for their complete analysis. To that end, hydride generation (HG), liquid chromatography (LC), gas chromatography (GC) and capillary electrophoresis (CE) have been used.[3] Once extracted in solution, the arsenic species can be separated using a suitable separation method. Arsenic species in aqueous extracts are separated by the conventional chromatographic and electrophorectic methods that, in most cases, can be directly connected to sensitive detectors. While capillary electrophoresis (CE) has exceptional separation capacities, because of very small sample capacity it lacks sufficiently low detection limits to be useful for routine environmental and biological analysis. On the other hand, gas chromatography is limited to the analysis of volatile analytes only.[8]

One of the most widely accepted methods for the determination of arsenic at trace concentrations in environmental and biological samples is hydride generation (HG). This method is based on the production of volatile arsines (Table 1.1) mainly of As(III), As(V), MA and DMA.[3,76] Production of arsines from arsenosugars has been reported recently.[77] Reaction of sodium borohydride and acid mixtures with the arsenicals to produce the respective arsines is almost universally accepted now after its introduction into analytical chemistry.[78] The volatile arsines are separated from the liquid reagents and transported to the detector by an inert gas. Two important advantages of the HG process are its ability to separate arsenic species from the sample matrix thus improving detection limit of most detection methods 100-fold, and the flexibility of the HG system to connect to various detectors (e.g., AAS, ICP-AES, ICP-MS) with or without the HPLC system.[3]

Arsenic in Plants: Extraction and Speciation

Since the arsenic compounds found in environmental and biological samples can be brought to solution, LC methods have been widely used for their separation.[3,8,26,27] The polymer-based Hamilton PRP- X100 anion-exchange column has been widely used with phosphate buffers for the separation of As(III), As(V), MA and DMA in solution. A number of arsenosugars have also been separated using the anion-exchange column.[3] The other arsenic compounds, TMAO, AB, AC and TETRA termed as cationic, are best separated on cation-exchange columns with pyridine-formate buffer solutions as the frequently-used mobile phases.[3,26,27]

There are a number of reports on the separation of both anionic and cationic arsenic species in a single run.[27] A record seventeen arsenicals in marine biota have been claimed to be separated simultaneously on an anion-exchange (Ionpak AS7 from Dionex) column with nitric acid eluents (pH 3.4 and pH 1.8) with a doubly-charged ion-pairing agent (0.05 mM benzene 1,2-disulfonate).[79] In this method, anions have been separated as usual while the cations have been retarded by their formation of negatively charged ion-pairs with benzene 1,2-disulfonate facilitating both anions and cations to be separated together in a single chromatographic run. Within a short time of the earlier report, another group claimed to separate twenty three organoarsenicals in marine samples using a cation-exchange (Ionospher-5C) column and aqueous pyridine-formate and gradient elution.[80] Since a limited number of arsenic standards was available and discrepancies were noted regarding matrix effects on the reported separations, these methods need to be further scrutinized.[27]

Fast separation of the anionic arsenicals (As(III), As(V), MA and DMA) has been accomplished by the use of narrow bore reversed-phase HPLC column with ion-pairing. Separation of these arsenic compounds was done in only 2 minutes.[81] The same species have also been separated on anion-exchange microbore column.[82] These microbore analyses have reduced sample and reagent consumptions. Since factors like time of separation, sample size and solvent consumption are becoming more important, greater use of the microbore columns is expected.[27] However, the resolution of the analytes in fast chromatography is found to have suffered from sample matrix-effects compared to the conventional chromatography with longer analyte retention time.[26]

1.7. ARSENIC IN PLANTS

The typical arsenic concentrations are well below 1 $\mu g.g^{-1}$ in plants grown on uncontaminated soils.[47,83] Depending on the levels of arsenic contamination, soil conditions and the species of plants, the concentrations of arsenic vary widely in plants and may be tens of $\mu g.g^{-1}$ in some plants grown on arsenic impacted soils.[84,85,86] High levels of arsenic are found in plants growing on many contaminated sites from gold mining activities.[19] In this study, the concentrations of arsenic in plants sampled from gold mining areas ranged from lower than 1 $\mu g.g^{-1}$ to greater than 500 $\mu g.g^{-1}$. Serious health concerns have arisen for millions of people due to consumption of foods derived from plants grown on arsenic contaminated soils.[49,50] Therefore, the analysis of arsenic in plants has become essential for risk assessments.

In early times, plants such as Douglas firs were analyzed for arsenic as a biogeochemical indicator of gold deposits.[6] Recently however, the determination of total arsenic (> 22,000 $\mu g.g^{-1}$) in arsenic-hyperaccumulating plant such as the Chinese brake fern (*Pteris vittata L.*) has received interest with the prospect of using the plant for arsenic remediation, a process known as phytoremediation.[87,88,89] The analytical methods commonly used for the determination of total arsenic in plants are the HG-AAS, ICP-AES and ICPMS.[3,8,27] Though it is an unconventional method, near-infrared spectroscopy (NIRS) has been shown to be a convenient time and cost saving method for the determination of total arsenic in plant tissues of the prostrate amaranth (*Amaranthus blitoides S. Watson*).[90] The total arsenic along with As(III) and As(V) in peach trees has been determined by a wavelength dispersive X-ray fluorescence method.[91] The X-ray fluorescence method with a detection limit 100 $\mu g.L^{-1}$ has been found to be quick and simple and was used for risk assessment.

1.7.1. Arsenic Species in Terrestrial Plants

Arsenic in Plants: Extraction and Speciation

Although many arsenic compounds present in the terrestrial and marine environments have been detected, As(V), As(III), MA and DMA are the predominant species found in terrestrial biota.[27] The predominance of inorganic arsenic, As(III) and As(V) in terrestrial plants due to anthropogenic activities has been demonstrated in several studies.[92] Vascular plants and bryophytes (mosses) from Yellowknife have been analyzed for total arsenic and water soluble arsenic species; the extracted arsenic was mostly inorganic. In some plants, small amounts of methylated arsenic species, MA and DMA, as well as arsenosugars were detected.[16] Lichens and terrestrial fungi of the Yellowknife area have been shown to contain higher arsenic than the background levels of the area. Both inorganic and organic arsenic species including AB, AC, TETRA were detected.[92]

Elevated arsenic levels in the surrounding biota due to natural exposure have been reported.[15] Arsenic levels in water from Meager Creek hot springs in British Columbia, Canada were shown to be naturally higher than the background levels. Plants, mushrooms and lichens of the Meager Creek area were analyzed for total arsenic and arsenic species. The major arsenic species extracted from these samples were As(V) and As(III) though small amounts of arsenosugars were also detected. However, more than 50% of the arsenic could not be extracted by using the standard 1:1 water-methanol extraction. Organoarsenic species, MA, DMA, TMAO, TETRA and trace amounts of AC have been identified in plants grown in soil above an ore vein.[93]

1.7.2. Extraction Methods for Arsenic in Plants

Arsenic must be extracted from solid samples and presented in an aqueous form for speciation by the more common methods such as the HPLC.[8,94] Many extraction methods have been explored because the extraction of arsenic from solid matrices is a critical step and high recoveries are required.[27,95] Mild but efficient conditions are needed for maximum extraction and, at the same time, to preserve the integrity of the chemical forms.[96,97] The diversity of arsenic and plant species in the environment as well as the locations of sampling complicate the determination of arsenic concentrations.[83,93] Moreover, since toxicity of arsenic is dependent on the chemical form as well as the concentration, both speciation and quantitative determination are needed.[26,98,99]

1.7.2.1. Solvent Extraction

Solvent extraction aided by physical shaking or sonication has been the traditional method for extracting arsenic from the solid matrices. Soxhlet and sonication are common tools used in the extraction of metals from plant matrices.[100,101] A number of extraction methods have been based on the mixture of methanol and water, a relatively mild extractant, aimed at maintaining the integrity of arsenic species in the samples.[27,97] Despite the common use of the methanol-water mixture for terrestrial plants, it has poor extraction efficiency. In their studies on arsenic in terrestrial plants from British Columbia and North West Territories of Canada, Koch et al. have shown that only about 50% of the arsenic could be extracted by using a 1:1 water-methanol and sonication method.[15,16,92] Other investigators, using 1:9 water-methanol, have extracted only 2.5-12% arsenic from plants grown on top of an ore vein[93] and, using 1:2 water-methanol solvent, extracted 73% arsenic from a certified reference material (GBW 82301-peach leaves).[95] In another study, only 6.3-16.1% arsenic could be extracted from plants of the Moira watershed (Canada) using 1:9 water-methanol extractions.[102]

Dilute (0.3 M) phosphoric acid was found to be a convenient extractant of arsenic species from terrestrial plants.[83] Combinations of acetonitrile-water[96,97] and methanol-water-chloroform[99] have been used as well. Extraction efficiency, according to the published reports, has varied widely depending on plant type and extractant.

Sonication is frequently used in combination with a solvent as a powerful tool of extraction, but the extraction efficiency has depended on the sample and its physical condition.[96,103,104] Pressurized liquid extraction (PLE), also known by the trade name ASE (Accelerated Solvent Extraction), is a rapid technique of solvent extraction at elevated pressure and temperature. Extraction efficiency with PLE was also found to be sample dependent: 80-102% for freeze-dried carrots,[105] 33% for rice[103] and 30%

for fresh plant materials[106] were reported. Microwave-assisted extractions were used for the analysis of arsenic from carrots grown in contaminated soil giving 46-69% extraction efficiencies.[107]

Treatment with chemical reagents such as acid[97,103] and base[97] has been used for the modification of sample matrices. Treatment with 2 M-tetrafluoroacetic acid for 6 hours yielded extraction efficiencies of up to 92% for rice samples.[103] NaOH performed better compared to the other reagents, but still with a low extraction efficiency of 36% maximum.[97] In another investigation,[108] 0.01 M-phosphoric acid was observed to stabilize the arsenic species in aqueous media.

Although many individual plant species and foods have been extracted using the various methods, there remains a need for more effective extraction methods that can be applied to terrestrial plants.[27] Very inefficient arsenic extraction by water-methanol from plants prompted other researchers to inquire for better methods of arsenic extraction.[102]

1.7.2.2. Enzyme and Gastric Fluid Extractions

Enzymes such as alpha-amylase (a cellulase) for cellulose have been used for the modification of the plant matrices for better extraction of analytes.[96,103,104] Treatment by the cellulase yielded an extraction efficiency of 104% for freeze-dried apples[96] and 59% for rice samples.[103] However, the enzymatic treatment did not improve the extraction efficiency in the seaweed (kelp) and the enzyme was mentioned to be expensive.[104]

Since the toxicity is directly dependent on the bioavailability of the element through the gastrointestinal tracts, the extraction efficiency of gastric fluid for the element is critically important. Historically, bioavailability of toxic elements has been carried out in vivo using the expensive and time-consuming method of ingesting soluble salts of elements toxic to animals or humans.[109] These experiments, however, did not take into account the physical and chemical characteristics of the toxicants in natural matrices that might affect their bioavailability. To overcome the limitations, in vitro extraction procedures mimicking the processes of the gastrointestinal tracts have been developed.[110,111,112]

Synthesized gastric juice has been used to determine bioavailability of metals including arsenic in the human stomach, and the extraction efficiencies are used to estimate the bioavailability of arsenic in both the human stomach and intestines.[110,111] GFE has been employed in a mass-balance and soil recapture technique for the determination of bioavailability of arsenic and other metals in soil matrices and it has been found that about 41% of the total arsenic was bioaccessible.[113] GFE extractions are currently being used in risk assessments of arsenic contaminated sites,[94] therefore, it is important to compare the extraction efficiency of the various extractants used in this thesis project to GFE extraction efficiency.

1.7.3. Arsenic in Horsetails (*Equisetum Sp.*)

Horsetails are known for many uses. The field horsetail (*E. arvense*) has long been valued for its therapeutically active soluble silicate acid.[114,115] It is used as a food, herb and vegetable,[116,117] and a source of silica for industry and technology.[118,119]

Horsetails, however, were popular among gold prospectors for a different reason. The prospectors used to believe that the horsetails accumulated gold on the auriferous ground. Though their belief was wrong, the horsetails grown on auriferous and arseniferous grounds were found to accumulate arsenic in high amounts and were tolerant to arsenic concentrations of 700 $\mu g.g^{-1}$.[120] There is a geochemical association of arsenic and gold; the horsetails, hence, were used as an indirect indicator of gold deposits. In another study, horsetails were found to be metal tolerant and were cited for their possible use in the phytoremediation of arsenic and mercury from contaminated soils.[121] Also in this study, the horsetails were found to accumulate the highest amounts of arsenic among all plant species tested. More than 500 $\mu g.g^{-1}$ arsenic was determined in the young field horsetail (see Chapter 4.10 for details). The study of this plant is important in the contexts of potential toxicity to humans and its potential to remediate contaminated soil.

1.8. THE RESEARCH OBJECTIVES OF PRESENT STUDY

Arsenic in Plants: Extraction and Speciation

The purpose of the present work was to develop more effective but simple methods for arsenic extraction from terrestrial plants, particularly for plants from contaminated sites. In order to be applicable for risk assessment, these methods need to be reliable and make use of readily available instrumentation. The other objectives of the study were to increase understanding of the causes of low extraction efficiencies attained by traditional water-methanol extractions, and to improve arsenic speciation, particularly the As(III), As(V), MA and DMA species, found in terrestrial plants grown on arsenic impacted sites.

In this study, a sequential method for arsenic extraction has been developed and successfully applied to extract and speciate arsenic from terrestrial plants. The predominant species of arsenic in plants, As(III) and As(V), have been studied for their assimilation by and extraction from different plant species.

The various schemes for the extraction of arsenic from plants have been proposed and evaluated in terms of extraction efficiency, species specificity, and sample handling efficiency. XANES experiments with fresh frozen, dry ground and residues of extraction have been conducted. Multi-element analyses in plant and soil samples have been carried out to determine correlation between arsenic and other elements.

CHAPTER 2: METHODS

2.1. SAMPLING AND ANALYSIS

2.1.1. Sample Collection and Preparation

At a preliminary trial of the study, plant samples were handpicked and stored in plastic bags overnight at 4°C. Plants were cut to remove roots from stems and washed with copious amounts of distilled water to remove the adhered soil and dust particles. Washed plants were air dried in the laboratory. The six plant samples of the first batch were ground using a laboratory plant grinder (Retsch, Brinkmann). About 100 grams of each plant sample was ground. The grinder was cleaned between samples by brush and paper towel. The ground plants were collected in plastic bags and stored in larger plastic bags in the freezer (-20°C). The samples were returned to room temperature before weighing for analysis and extraction.

In the latter expeditions for sample collection, in order to avoid soil contamination, all plant samples were gathered by cutting above the roots using small garden shears or scissors and were put into Ziploc™ bags and stored at 4°C overnight. The same plant specimen sampled from a location, usually a square meter area, were mixed together, a portion was washed with copious amounts of distilled water and let drip on fresh paper towels. After the initial draining, the samples were placed on brown paper spread on top of tables inside the laboratory to air dry. The remnants of the fresh plants were stored in the freezer for future reference.

Because of time consuming process using the heavy-duty grinder, and the large number of samples collected in Deloro, a small coffee grinder was dedicated and employed for subsequent trips. The appropriate amounts of the air-dried plant samples (approximately 20-30 g) were taken in 100 mL glass beakers and placed in the oven at about 70°C overnight to remove more moisture from the plants. After removal from the oven and cooling, the plant samples were immediately sealed to protect them from ambient moisture. This treatment rendered the plant samples crisp and easy for grinding. The plastic lid of the grinder was washed to remove all adhering fine particles and was dried before re-use while the metallic grinding cup was cleaned thoroughly between samples using brush and moist paper towel, if needed. All ground samples were collected in plastic vials and stored at - 20°C. As with previous samples, all frozen samples were brought to room temperature prior to weighing for analysis. The details of sample collection from the Deloro site specifically are given in Chapter 4, Section 4.3.

2.1.2. Plants from Yellowknife

Plants from Yellowknife in NWT, Canada used in this study were collected from Con and Giant gold mining areas in September 2000 and identified by the Environmental Sciences Group (ESG) of Royal Military College of Canada at Kingston, Ontario. Whole plants or plant shoots were handpicked and kept cool in Ziploc™ bags until processing in the laboratory. Plants were washed thoroughly with copious amounts of water and Sparkleen™ solution to remove the soil and other particles, and finally rinsed with deionized water. Roots were separated from the shoots and frozen until processed for analysis.[16]

The frozen plant samples received from the ESG were defrosted and laid on the tables to air dry. The plants were ground by the laboratory grinder and stored in the freezer in plastic bags. The ground and frozen plant samples were allowed to return to room temperature before sampling for analysis.

2.1.3. Preparation of Soil Samples

Surface (0 – 10 cm) soil samples were gathered in Whirlpaks® (bags) using individual sampling scoops (Scienceware® Sterile Sampling system). All soil samples were stored at 4°C until air dried in the laboratory. The soil samples were prepared for analysis based on the USEPA Method 200.7 with small modifications (that is, sampled without sieving and using smaller sample weights). Aliquots of 125-150 g from the representative samples were air-dried and homogenized by clean mortars and pestles discarding the hard and large pebbles. Aliquot portions of the finely powdered soil samples were analyzed for total arsenic and the remainders were stored in plastic bags at room temperature.

2.1.4. Digestion of Plant Samples

0.5 g (± 0.1 mg) portions of the ground plant samples were weighed in Vycor® glass crucibles and ashed in a (Fisher Scientific ISOTEMP programmable) muffled furnace with the temperature ramps: 150°C for 20 minutes, 250°C for 60 minutes, 500°C for 3 hours and return to room temperature. The ashed plant samples and method blanks (crucibles without samples) were digested with 4 mL 1:3 v/v HNO_3/HCl solutions for 4 hours on a hot plate under watch glass covers. After removing the watch glass covers, 2-3 drops of 50% H_2O_2 were added and the volumes reduced to approximately 2 mL. All samples were then transferred to graduated tubes and made up to 12.5 mL volumes with DDW and were then filtered (Fisherbrand® Quantitative, medium porosity).

2.1.5. Digestion of Soil Samples

Dried soil samples (0.5 g ± 0.1 mg) were weighed in the acid-cleaned graduated digestion tubes and mixed with 8 mL 1:3 v/v concentrated HNO_3/HCl solution. The soil samples were allowed to presoak in the acid for an hour under the fume hood. Then the tubes were placed, a few at a time, in the preheated (to 200°C) aluminum block on the hot plate and kept under close observation. Both cold-water bath and vortex were used to keep samples from overflowing due to vigorous reaction and foaming. Once settled, the samples were allowed to digest overnight and to reduce to 0.5-1.0 mL. After cooling in a rack, digested residues were made up to 25 mL with DDW, vortexed to mix and filtered. The method blanks comprising the acid reagents were heated with a few glass chips (broken tips of Pasteur pipettes) to avoid bumping, and treated in the same manner. MESS-3 reference standard and duplicate samples were analyzed for quality assurance.

2.2. ANALYTICAL TECHNIQUES USED

Both ICPAES and HGAAS were used for total arsenic determinations in the acid digests of the solid plant matrices as well as in the liquid plant extracts. Estimates of the low-level arsenic concentrations were obtained from the ICP analysis prior to employing the more sensitive hydride generation method because of the small linear range (5-30 $\mu g.L^{-1}$) of the HGAAS. HPLC-HGAAS was used for the speciation of the arsenic compounds in the extracts. A number of the plant extracts were analyzed by HPLC-ICPMS as a quality assurance measure. The methods of plant extraction as developed in this study will be discussed in the respective chapters.

2.2.1. Total Arsenic by ICP-AES

The ICPAES used in this study had the advantages of simultaneous multi-element determination and analysis without chemical pretreatment necessary for the hydride generation method for arsenic determination. The ICP analyses were conducted by a Varian VISTA AX CCD Simultaneous ICP-AES instrument. The central feature of the axially viewed ICP-AES is the Charge Coupled Device (CCD) detector with 70,000-pixel arrangement stabilized at - 35°C intended to analyze low concentration samples. This pixel arrangement matches the two-dimensional image from the argon purged high-resolution echelle polychromator giving continuous wavelength coverage from 167-785 nm and facilitating the elimination of spectral interferences.

The acid digested and filtered sample solution was introduced to the RF induced argon plasma as a uniformly distributed aerosol by the use of a concentric pneumatic glass nebulizer and argon as the carrier gas. A dynamic range of analyte concentrations from parts per billion (ppb) to percent levels could be determined with nominal or no dilution by the careful selection of emitted wavelengths. However, the range for each element varied and depended on the intensity of the chosen wavelength. As many as 72 elements could be simultaneously determined by this method. The wavelengths used for arsenic determinations by ICP-AES were arsenic-188.980 nm and arsenic-193.696 nm and the averages of concentrations from two wavelengths were used.

The total arsenic was determined by ICP-AES using scandium/indium as internal standards following the standard ASU method (Analytical Methods, ASU, Queen's University, Kingston, ON, Canada) for arsenic in plants with a method detection limit 15 $\mu g.L^{-1}$ in solution. Reagent blanks as well as initial and continuous calibration verifications (ICV and CCV) were used as Quality Control/Quality Assurance (QA/QC) measures. The ICV is a standard solution made from different stock solutions than the calibration standards and is used to verify the calibration standard levels. The CCV is used to verify excessive instrument drift and is used every time after a certain number (usually 16 or less) of samples analyzed. Multi-element mixed calibration standards ranged from 0.10-5.0 $mg.L^{-1}$ and were prepared in 2 M-HNO_3 from commercial ICP element standards. More on calibration has been described in QA/QC in Section 2.3. All instrument parameters for arsenic determination are listed in Table 2.1.

2.2.2. Total Arsenic by HGAASS

Because of the lower arsenic detection limits of HGAAS than those of ICPAES (MDL <0.19 $\mu g.L^{-1}$ for HGAAS vs. 15 $\mu g.L^{-1}$ for ICPAES in solution), HGAAS at the ASU was employed to analyze arsenic in very low concentration samples. The HGAAS analysis for total arsenic was carried out with an AA/AE Spectrophotometer (Instrumentation Laboratory of Allied Analytical Systems, USA) equipped with a vapour generation accessory (Varian VGA-76) for hydride generation and separation, and atomization was performed in an air-acetylene flame. The arsenic AAS was conducted at 197.2 nm wavelength and 1.0 nm bandwidth. In the prereduction step, appropriate volumes of 20% w/v hydroxylamine hydrochloride (2.5 mL for 25 mL final volume), 15% w/v potassium iodide (1.67 mL for 25 mL final volume) and concentrated hydrochloric acid (2 mL for 25 mL final volume) were added to measured aliquots of sample extracts. Since As(III) has the highest HGAAS sensitivity, all As(V) was reduced to As(III) by KI at low pH condition during the prereduction step.

Table 2.1. Instrument parameters for arsenic determination.

ICPAES		HPLC-HGAAS	
Nebulizer pressure	200 kPa	No. of re-sample	99
rf power	1.2 kW	Measurement time	3.3 secs
Plasma flow	15.0 L.min^{-1}	Wavelength	193.7
Auxiliary flow	1.50 L.min^{-1}	Lamp current	75%
Measurement time	5.0 secs	Bandpass	1.0 nm
Replicates	4	Signal	Continuous
Replicate time	5.0 secs	Backgr. correction	D2 Quadline
Internal standards	Sc/In, 50 ppm	Technique	Vapour
HGAAS		Vapour mode	Electrical heating
NaBH$_4$ conc.	0.6% in 0.5% NaOH	Furnace temp.	900° C
HCl conc.	10 M	NaBH$_4$ conc.	1% in 0.1% NaOH solution
NaBH$_4$ flow	1.0 mL.min^{-1}	HCl conc.	1 M
HCl flow	1.3 mL.min^{-1}	NaBH$_4$ flow	4.0 mL.min^{-1}
Sample flow	6.0 mL.min^{-1}	HCl flow	2.3 mL.min^{-1}
Atomizer	Air-Acetylene flame	Carrier gas	Argon
Carrier gas	Nitrogen	**HPLC**	
Lamp	Arsenic 197.2 nm	Anionic column	Hamilton PRP X100, 250x4.6 mm
Band width	1.0 nm	Guard column	Hamilton PRP X100, 20x4.6 mm
ICPMS		Mobile phase	20 mM (NH$_4$)$_3$PO$_4$, pH 6.0
rf power	Forward: 1382 W	Cationic column	Varian Chromsep 150x4.5 mm,
	Reflected: <1 W		ionosphere 5C
Ar flow rate	Coolant: 13 L.min^{-1}	Mobile phase	20 mM pyridinium formate, pH 2.7
	Nebulizer: 0.93 L.min^{-1}	Injection volume	100 μL (HGAAS), 10 μL (ICPMS)
	Auxiliary: 0.83 L.min^{-1}	Flow rate	1.5 mL.min^{-1} (HGAAS)
Dual time	300 ms		1.0 mL.min^{-1} (ICPMS)
Internal standards	Rh (for anions), In (for cations), 5 μg.L^{-1}	Mode	Isocratic

Reagent blanks and calibration standards along with the QC standard were prepared at the same time following an analogous procedure using commercial elemental standards for ICP/AAS. The linear arsenic calibration range of HGAAS analysis was 5-30 μg.L^{-1}. The mixtures were allowed to react for approximately 1 hour and then pumped (6 mL.min^{-1}) into the HGAAS (air-acetylene) flame along with the borohydride (0.6% w/v NaBH$_4$ in 0.5% w/v NaOH, 1.0 mL.min^{-1}) and 10 M-HCl solution (1.3

mL.min^{-1}) by the hydride generation system. The method detection limit (MDL) of the HGAAS was calculated to be 0.18 µg.L^{-1}. The detailed QA/QC is presented in Section 2.3. The determination of reliable detection limit (RDL) of As(III), As(V), MA and DMA, and the method detection limit (MDL) of ICP-AES and HG-AAS are given in Chapter 3 (Section 3.10).

2.2.3. Arsenic Speciation by HPLC-HGAAS

All HPLC-HGAAS analyses were carried out at the ESG Laboratory. Standards and plant extracts were introduced in the 100-µL sample loop using 1-mL syringes fitted with the syringe filters (Millipore Milex-HV Hydrophilic PVDF 0.45 µm) and injected into the HPLC column. The HPLC column was an anion exchange Hamilton PRP-X100 (250 mm x 4.6 mm) dimension and was fitted with a guard column of the same material (30 mm x 4.6 mm). The carrier solution was ammonium phosphate buffer (20 mM-H_3PO_4, pH adjusted to 6.0 by titrating with NH_3 solution) flowing at 1.5 mL.min^{-1}. The detection of arsenic was carried out at 193.70 nm wavelength by a SOLAAR 969 AAS instrument. The AAS instrument was equipped with the VP90 vapour generation accessory, an EC90 furnace and quartz T-tube and was operated using SOLAAR software (all supplied by ThermoInstruments, Canada).

Hydride generation was achieved by mixing the eluent from the HPLC with 1 M-HCl and 1% $NaBH_4$ stabilized in 0.1% NaOH. The arsines were separated from the liquid by introduction into a VP90 vapor generation unit connected to a SOLAAR 969 AAS instrument as described above. HPLC chromatograms were generated from AAS data by using MS Excel.

Mixed standards of As(III), As(V), MA and DMA were prepared from concentrated stocks in 50 µg.L^{-1}, 100 µg.L^{-1}, 200 µg.L^{-1} and 500 µg.L^{-1} concentrations by mixing appropriate volumes in DDW. External calibration curves were obtained by plotting peak areas against concentrations. Besides mixed standards, independent standards were also run as a QA measure.

2.2.4. Arsenic Speciation by HPLC-ICPMS

HPLC-ICPMS experiments were performed as quality assurance of the earlier results of As(III), As(V), DMA and MA from HPLC-HGAAS. Since not all arsenic compounds (e.g. AB, TMAO, Tetra) were analyzed by HPLC-HGAAS, plant extracts were further analyzed in HPLC-ICPMS using a cation exchange column to ascertain presence of organoarsenic compounds other than MA and DMA that were already determined by HPLC-HGAAS in the sample extracts. The cation exchange column was a Chrompack Ionosphere C column (3mm x 100 mm, 5-μm particles) and used 20 mM pyridine formate (pH 2.6) mobile phase at 1.0 mL.min^{-1} flow rate. The HPLC was connected to a PQ ExCell ICP-MS and operated in normal mode.

2.2.5. X-Ray Absorption Near Edge Structure (XANES) of Arsenic Species - Fingerprints from XANES

XANES spectra of 19 plant samples (Table 5.3) containing arsenic levels above the XANES detection limit approximately 10 μg.g^{-1} were collected at the bending magnet beam line of Pacific Northwest Consortium Collaborative Access Team (PNC-CAT), Sector 20, at the Advanced Photon Source, Argonne National Laboratory in Illinois.[122] Subsamples were loaded into a stainless steel sample holder consisting of a plate with 3 x 10 mm holes in it that had one side of the plate covered in a plastic tape that is permeable to X-rays (Kapton™). The other side of the plate was covered in Kapton™ tape, sequestering each sample between two sheets of the tape, which allowed the X-ray beam to pass directly through the sample.

The plate containing the samples was positioned in the path of the X-ray beam, at a 45° angle to both the beam and the detector (the latter was positioned 90° to the beam). A silicon (111) double-crystal monochromator (resolution ~ 5000 E/dE at 12 keV, detuned to 85% of maximum at 12,100 eV) and a rhodium-coated harmonic rejection mirror provided X-rays for measurement. The monochromator was calibrated using the gold L3 absorption edge (11,919.7 eV)[123] for measurements at the arsenic K-edge

(11,868 eV). A slit of 1 mm vertical by 4 mm horizontal was used after the mirror to reduce scatter from different components in the beam line. Nitrogen-filled transmission ionization chambers were present before and after the samples for normalization to the incident intensity and transmission measurements, respectively. Fluorescence data were collected using a solid-state Ge(Li) detector (Canberra model GL0055PS) or an argon-filled fluorescence ionization chamber[124] for standards (pure KH_2AsO_4 and As_2O_3 powders, reagent grade, Fluka). Typically, five scans were collected and averaged before background-removal and normalization-to-edge-jump. The WinXAS program[125] was used for processing the spectra.

XANES is the x-ray absorption spectrum of an element within about 50 eV of the absorption edge. XANES is strongly sensitive to the local coordination chemistry, oxidation states and bonding characteristics of the absorbing element. The speciation of element in a sample is possible because many elements show significant edge (i.e., binding energy) shifts with oxidation state. As(III) and As(V), the predominant arsenic species in plants, absorb at 11.8713 and 11.8753 keV, respectively.

XANES experiments were conducted to determine the chemical speciation of arsenic in the Yellowknife and Deloro plant matrices. Plant samples comprising fresh frozen, dry ground, and dry residues after extractions, were sent to the Advanced Photon Source, Argonne National Laboratory in Illinois. Once received, the XANES data were analyzed by using WinXAS programme files.

2.3. QUALITY ASSURANCE AND QUALITY CONTROL (QA/QC)

In addition to QC described under different analytical methods, the sample preparations and analytical procedures were accompanied by blanks and spiked standards. Samples were analyzed in double or triple independent replicate to ascertain reproducibility of the analytical results. The number of duplicates and blanks in each analysis ranged 15-25% of the samples analyzed. Sample and spike concentrations were calculated using linear equations obtained from the external calibration curves. The total amount of arsenic in a number of reference materials [CRM GBW 07603 Bush Branches and

Leaves, 1.25 µg.g^{-1} total As; SRM NIST 1575 Pine Needles, 0.21 µg.g^{-1} total As; and IAEA-140/TM Seaweed (Fucus sp.), 42.2-46.4 µg.g^{-1} total As] were determined using duplicate or triplicate independent samples. The results of the reference material analysis are reported in Table 2.2. The four-point linear calibration range of ICPAES was 0.10-5.0 µg.g^{-1}. Except for the concentrations close to the LOD, results were within 10% of the certified values.

The results of total arsenic in a number of plants determined by the ashing and concentrated HNO$_3$ acid digestion methods[63] were compared. The concentrations of arsenic of the CRM by the wet method ranged within 90-100% of the certified results and no significant difference was found between the results from the two methods. However, the ASU method that removed plant organic materials by dry ashing was more convenient to follow in this lab than the wet HNO$_3$ method. Mass balance experiments comprising determinations of arsenic in the extracts and plant residues accounted for the total arsenic (100% recovery) showed no loss or gain of arsenic during analyses involving several steps (Table 3.4 and Table 3.5).

Table 2.2. Determination of arsenic in certified reference materials: seaweed, bush branches and leaves, and citrus leaves by nitric acid digestion and ICP-AES detection.

Sample ID	Results of analysis (Mean ± ave. deviation), µg.g^{-1}	Certified values, µg.g^{-1}
FUCUS (Seaweed), n = 3	39.9 ± 0.5	42.2 - 46.4
CRM (Bush branches and leaves), n = 2	1.25 ± 0.01	1.25 ± 0.24
NIST (Citrus leaves), n = 2	0.19 ± 0.04	0.21 ± 0.04

# CHAPTER 3: ARSENIC EXTRACTION AND SPECIATION IN PLANTS GROWN ON ARSENIC POLLUTED SOILS - DEVELOPMENT OF SEQUENTIAL EXTRACTION METHOD

3.1. INTRODUCTION

The environmental contamination of arsenic due to natural and anthropogenic activities has been discussed in Chapter 1. The specific toxicity of arsenic is well documented and recognized as a global health problem, which is also discussed in Chapter 1. In order to determine the health risks of arsenic to humans and wildlife, its characterization is necessary and it must be efficiently extracted from plant matrices.

The arsenic extraction methods described in Chapter 1 used complicated instruments and methods (e.g., ASE, microwave extraction and supercritical fluid extraction) involving elevated pressure and temperature, and special reagents. An efficient but straightforward and easy to use arsenic extraction method for extracting a large number of environmental samples was lacking.[27,102] The goal of the current work was to develop more effective but simple methods to extract arsenic from terrestrial plants, particularly for plants from sites with high levels of arsenic contamination. In order to be applicable to risk assessment concerning samples from arsenic impacted sites, these methods need to be reliable and use readily available instrumentation. Water, methanol, 1:1 v/v water-methanol, 0.1 M-NaOH and dilute HCl at various concentrations, and both solvent-sonication and solvent-Soxhlet methods were explored. These solvents have been used for extracting arsenic from fresh and freeze-dried plants, fruits and vegetables, and biological samples with varying successes as are described in Chapter 1. The traditional Soxhlet extraction method and more effective sonication technique are also described.

Arsenic in Plants: Extraction and Speciation

Many experiments were carried out for the development of an efficient arsenic extraction method. The results of those experiments and the subsequent development of the sequential arsenic extraction method are discussed in this chapter.

3.2. TOTAL ARSENIC IN SOIL AND PLANT SAMPLES

The methods of determination of total arsenic in plants and soil samples are described in Chapter 2. Six plant samples were collected from 4 sites at Deloro as described in Chapter 4. These samples were used to develop the analytical methods. Results of total arsenic in the 4 soil samples are listed in Table 3.1 and for the 6 plant samples in Table 3.2. In this work, ICP-AES and HG-AAS served as tools for cross-comparison of the analytical results.

Table 3.1. Arsenic in soil samples collected from arsenic waste tailings and abandoned goldmine areas at Deloro.

Soil sample	Soil arsenic concentration, $\mu g \cdot g^{-1}$	Corresponding plant sample
1	100,000	P-1, P-2, P-3
2	335	P-4
3	496	P-5
4	2710	P-6

P-1: Field horsetail (*Equisetum arvense*), P-2: Cattail (*Typha sp.*) and P-3: Alkali grass (*Puccinellia sp.*), P-4: Bracken fern (*Pteridium aquilinum*), P-5 and P-6: Smooth horsetail (*Equisetum laevigatum*).

Results of total arsenic in the plants from both methods agree well. The RSD values of the average (n = 3) total arsenic content of the plants ranged from 2.2% to 11% and from 2.0% to 7.7% determined by ICP-AES and HG-AAS, respectively. A better precision from HG-AAS was expected for the lower concentrations since HG-AAS had a lower detection limit for As than ICP-AES.

Table 3.2. Total arsenic determined by ICP-AES and HG-AAS. Results are mean ± 1 s.d. from 3 independent determinations.

Plant sample ID	Total As by ICP-AES, $\mu g.g^{-1}$	Total As by HG-AAS, $\mu g.g^{-1}$
P-1	241 ± 9	234 ± 11
P-2	14.6 ± 0.7	14.6 ± 0.2
P-3	146 ± 5	147 ± 3
P-4	4.03 ± 0.45	3.78 ± 0.20
P-5	2.25 ± 0.05	1.83 ± 0.14
P-6	24.4 ± 0.7	23.7 ± 0.6

The three plants with higher arsenic content were selected for further extraction experiments: P-1, field horsetail (241 $\mu g.g^{-1}$ total As), P-3, alkali grass (146 $\mu g.g^{-1}$ total As) and P-6, smooth horsetail (24.4 $\mu g.g^{-1}$ total As). Representative pictures of plants and Deloro sites are shown in Photos 3.1-3.4.

3.2.1. Total Arsenic by Dry-Ashing and Wet-Ashing Methods

The total arsenic concentrations in plant and soil samples were determined following standard ASU methods as described in the methods section. In the ASU method, plants were ashed in a temperature-programmed furnace prior to acid digestion. There was some concern regarding loss of volatile arsenic compounds, if there were any, because of the ashing process in the oven. To attend to this question, wet acid digestion without prior ashing was applied to a number of plant samples and the results were compared with those obtained from ASU method.

Six plant samples were analyzed by concentrated HNO_3 acid wet digestion following a method published earlier.[63] The digested samples were analyzed by ICP-AES for total arsenic. The total arsenic results of the direct wet digestion and dry-ashing prior to 1:3 v/v HNO_3/HCl digestion methods are presented in Table 3.3. The mean arsenic concentrations from the two methods appeared to agree well and a means comparison by Student's t test for paired samples showed no significant difference (see also Table 2.2).

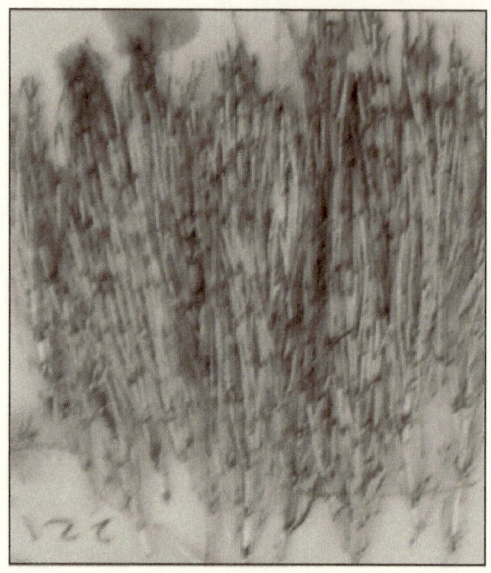

Photo 3.1. A sample of Field horsetail (*Equisetum arvense*).

Photo 3.2. A sample of Smooth horsetail (*Equisetum laevigatum*).

Photo 3.3. The remnants and reminders of 100 years' of gold mining activities at Deloro.

Photo 3.4. Arsenic tailing pond near sampling sites 4, 5 and 6.

Table 3.3. Total arsenic determined by wet ashing by HNO_3 (HNO_3 Method) and dry ashing in oven and digesting by 1:3 v/v HNO_3/HCl (ASU method).

Sample ID[a]	HNO_3 Method, $\mu g.g^{-1}$	ASU Method[b], $\mu g.g^{-1}$
DL-101	418 ± 12	410
DL-108	11.7 ± 0.4	11.3
DL-115	9.95 ± 0.13	9.1
DL-201	46.9 ± 0.8	48.4
DL-209	22.0 ± 0.5	20.3
DL-212	46.6 ± 1.4	42.0

[a] All samples were analyzed in duplicate except DL-209 in triplicate. Results are Mean ± ave. deviation from 2 or 3 independent determinations.
[b] Results are from total arsenic determination of Deloro plants.

The above experiment showed that there was no significant loss of arsenic due to ashing in the muffle furnace prior to acid digestion by the ASU method. The dry ashing method removed the organic plant materials that, when present, interfered in the wet acid method in sample handling and analyzing. As well, the ASU method was easy to follow in the standard ASU setup.

3.3. EXTRACTION OF ARSENIC FROM PLANTS: SOXHLET AND SONICATION PROCESSES

3.3.1. Solvent-Soxhlet Extraction Method

Three plants were extracted in a total of 100 mL solvent by the Soxhlet method which is routinely used to extract solid materials (EPA Method 3540B). Portions of the ground plant samples (0.5 g or 0.25 g) were accurately (± 0.1 mg) weighed into the cellulose thimbles (Fisherbrand® 25 mL). Water (DDW), methanol and 1:1 water-methanol were used as solvents. Approximately 60-70 mL of solvent was added to the round bottom distillation flask and heated gradually to its boiling point on ceramic or electric coil heaters. Soxhlet extraction lasted for six hours (4 cycles per hour) for each batch except for the exhaustive extraction. After the distillation process was over, extracts were transferred to 100 mL volumetric flasks and the volumes were made up to the marks with the respective solvents used for distillation.

3.3.2. Solvent-Sonication Extraction Method

A schematic representation of the solvent-sonication extraction process is given in Figure 3.1. The solvents used to extract arsenic from plant matrices were water (DDW), methanol, 1:1 v/v water/methanol, 0.1 M -, 0.05 M -, 0.02 M- and 0.01 M-HCl solutions, and 0.1 M-NaOH. Initially 0.5 g powdered sample was accurately ($\pm$ 0.1 mg) weighed into 15 mL extraction tubes and 10 mL of extractant was introduced to the samples in the extraction tubes. After closing the lids securely, sample and extractant were mixed thoroughly by a vortex apparatus for one minute. The samples were then sonicated for 20 minutes in the first step of extraction. The resulting mixtures were centrifuged at 3000 rpm for 10 minutes for all extraction steps. In the subsequent steps of extraction, samples that settled at the bottom by centrifugation were dispersed in the extractant solvent by tapping, shaking and vortexing, and were sonicated for 10 minutes. Extraction of samples was carried out five times (5 steps) initially, and then three times (3 steps) after the optimization of the process. The supernatant solutions after each centrifugation were decanted in 100 mL plastic bottles or 32 mL clear glass vials with Teflon lined lids. The extracts were filtered before analysis.

Initially, a roto-evaporation technique was used to remove methanol from the 1:1 water-methanol extracts. Reduced pressures of ~200 mbar for methanol to ~110 mbar for water were employed at 45-50°C flask temperature. The 30 mL or 50 mL extracts were reduced to about 2-3 mL. The reduced extracts were regenerated with DDW to the appropriate volumes. In the preliminary ICPAES experiments, the recovered concentrations of organoarsenic spikes (MA and DMA) in the roto-evaporated extracts were found to be significantly higher than the introduced amounts. The reason for high values was not clear; however, the presence of small amounts of methanol in the extracts after roto-evaporation may have been responsible. Complete evaporation in a roto-flask was not practical due to longer time of evaporation and regeneration of the extracts one at a time.

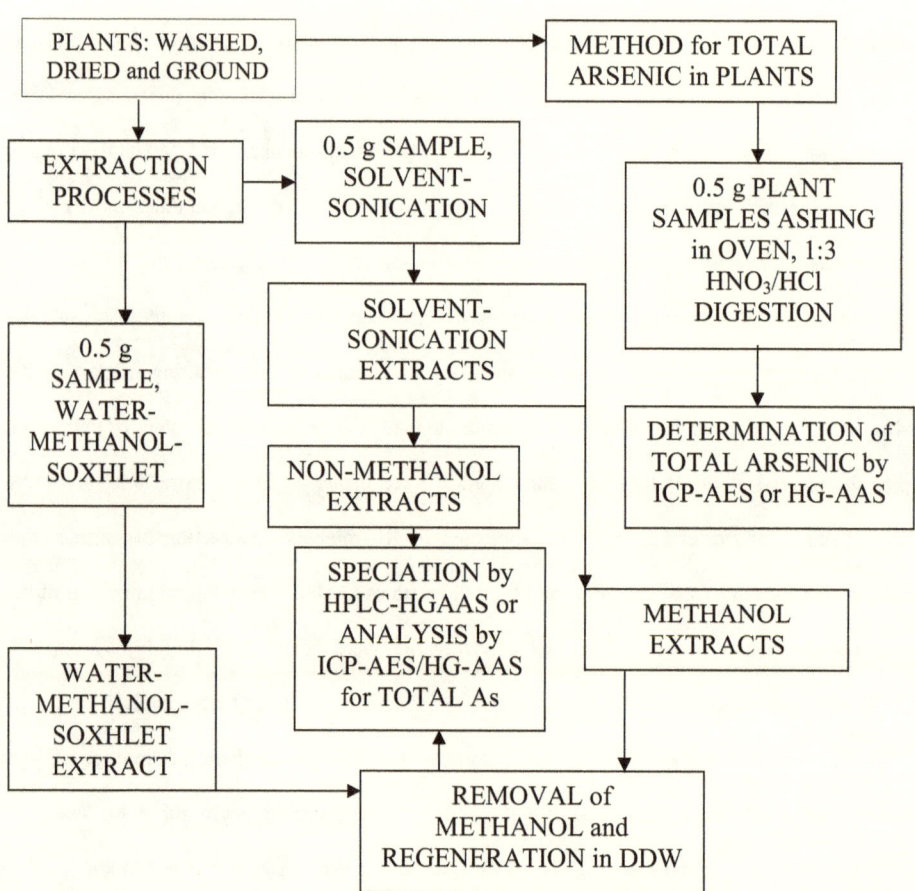

Figure 3.1. Flowchart showing initial methods of arsenic extraction and total arsenic determination.

Arsenic in Plants: Extraction and Speciation

To remove alcohol completely from the extracts, the sample vials with their lids open were placed in a sample rack and dried in an oven at a low temperature (50°-60°C) usually overnight or during weekends. The samples were regenerated in distilled deionized water (DDW) and analyzed by ICPAES and the spike results were found to be within ± 10% of the introduced concentrations. The procedure also turned out to be very convenient in terms of handling the extracts in large numbers and reducing the analysis time. Due to the use of the disposable sample vials and drying and regenerating each sample separately, the risks of carryover errors of analyte concentrations were eliminated. While many samples can be simultaneously dried and regenerated in an oven and on a shaker, respectively, roto-evaporation must be carried out individually for each sample and be constantly monitored.

3.3.3. Soxhlet and Sonication Processes: Results and Discussion

The extraction efficiency (EE) of a particular solvent is defined as the percentage of total arsenic extracted by the solvent from a plant. The results of extraction experiments are shown in Figure 3.2 and Figure 3.3 for Soxhlet and sonication experiments, respectively. Of the three solvents tested in the 6-hour Soxhlet extractions, water extracted the most arsenic from the plant samples (Fig. 3.2). The water-Soxhlet EE

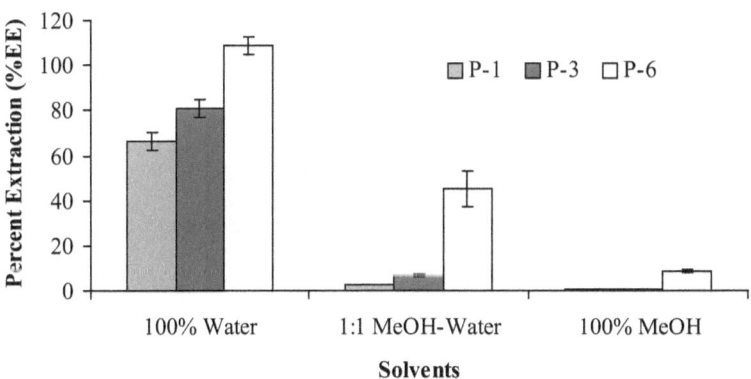

Figure 3.2. Extraction efficiencies (EE) of different solvents in solvent-Soxhlet method for three plants are presented. Error bars are ± 1 s.d. from three independent determinations.

for arsenic from plants P-1, P-2 and P-3 were 66%, 81% and 110%, respectively. In contrast, 100% methanol showed EE of 0.5%, 0.6% and 9% for P-1, P-3 and P-6, respectively.

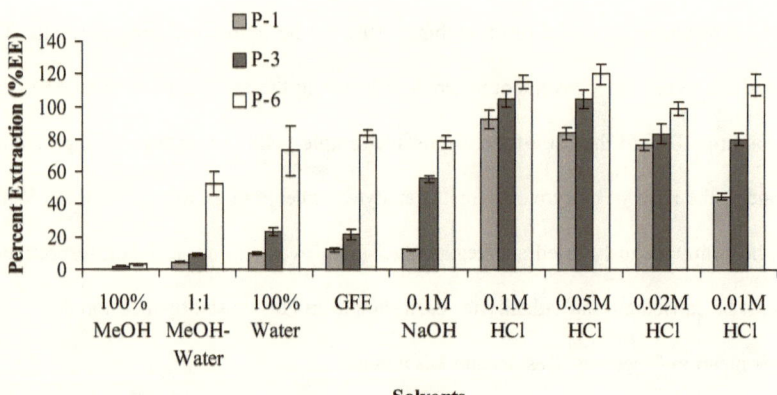

Figure 3.3. Extraction efficiencies (EE) of different solvents in solvent-sonication method for three plants are shown. Error bars represent ± 1 s.d. from three independent determinations.

Figure 3.3 illustrates extraction efficiencies for solvent sonication extractions using water (DDW), methanol, 1:1 water/methanol, 0.1 M-, 0.05 M-, 0.02 M- and 0.01 M-HCl solutions, and 0.1 M-NaOH. Water showed EE of 10%, 23%, and 74% for P-1, P-3, and P-6, respectively. Water-Soxhlet was more effective than water-sonication because water-Soxhlet is an exhaustive extraction technique and carried out at a higher temperature. Very low efficiency was obtained by methanol extraction in both Soxhlet and sonication methods. The efficiency of exhaustive Soxhlet extraction was determined by extracting P-6 (n = 3) for 2, 4, 6, and 12 hours with water as solvent. A plot of Soxhlet extraction efficiency versus time of extraction is presented in Figure 3.4. The results showed that a plateau was reached at around six hours of extraction by the Soxhlet. However, different plants have different content of total arsenic and physiology to deal with the toxic element. The plant matrices also vary with species making it difficult to generalize any conclusion regarding extraction of arsenic from plants. Besides, the Soxhlet method is time consuming, delicate in the setup and slow in sample handling.

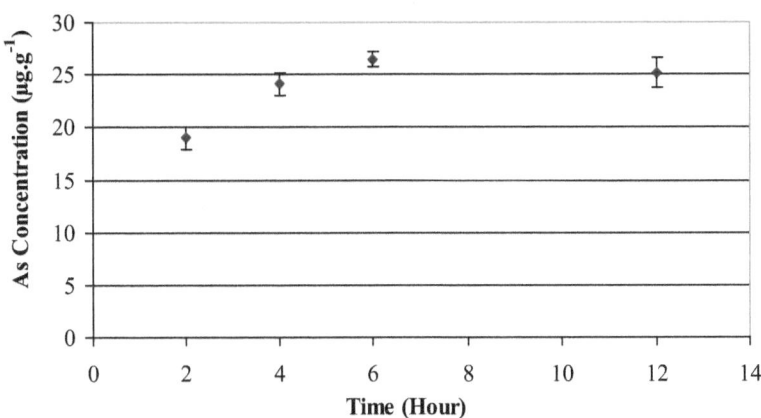

Figure 3.4. Results of exhaustive extraction of arsenic from plant (P-6) by water-Soxhlet process are presented. Error bars represent ± 1 s.d. from 3 independent determinations.

NaOH solution aided by a rotary mixer has been shown to improve extraction of As from plant materials: 22%, 32% and 36% from freeze dried poplar leaves, pine shoots and spruce shoots, respectively.[97] In our study, NaOH performed similarly to water and brought some improvement, from 23% to 56%, in the extraction of As in P-2 (Fig. 3.3), but, did not improve EE for the other two plants (P-1 and P-6). This is possibly due to the differences in the plant species and their matrices.

Introduction of HCl induced a marked improvement in the EE's of arsenic for all plant samples. Increases from 10% to 93% (P-1), from 23% to 105% (P-3) and from 74% to 117% (P-6) were observed by switching from water to 0.1 M-HCl as the extracting medium (Fig. 3.3). If the plants were extracted with 1:1 water-methanol alone, EE's of 4.3%, 8.8% and 54% for P-1, P-3, and P-6, respectively, would have been achieved. These EE's, undoubtedly, are much lower than those accomplished by the extraction with 0.1 M or 0.05 M HCl. Performances of 0.05 M-HCl and 0.1 M-HCl were similar, but did not improve significantly with higher HCl concentration as tested (see Table 3.14). This observation is important in terms of speciation processes where milder extraction medium is more desirable. Although concentrated HCl was used to determine total inorganic arsenic in biological and plant samples[96,126,127] mild HCl has not been used for the extraction and speciation of arsenic in plants grown on highly contaminated soils.

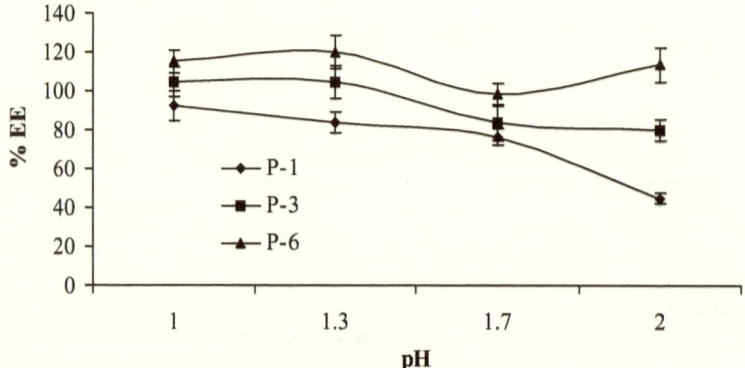

Figure 3.5. Extraction efficiency of plant as function of solvent pH is shown. Error bars represent ± 1 s.d. from 3 independent determinations.

The effect of HCl concentration (pH) on the EE was evaluated by comparing the extraction results of a number of plants extracted by 0.01 M-, 0.02 M-, 0.05 M- and 0.1 M-HCl. Figure 3.5 shows graphic representation of EE as function of pH. For plant P-1, EE increased from 44.8% to 92.4% with a decrease of pH from 2 to 1. For P-3, it improved from 80.1% to 104%, and in case of P-6, EE remained on the average above 100%. This result showed that at higher pH, the range of EE widened and became dependent on plant species.

3.4. THE EFFECT OF ARSENIC CONCENTRATION ON EXTRACTION EFFICIENCY (EE)

The total As content in plants varied widely and appeared to have some influence on the EE. In this study, it was observed that EE was inversely related to the total arsenic content of the plant samples. Our published results from solvent-Soxhlet extraction showed (Figure 3.6) decreasing EE with increasing total As content of plants.[94] A similar trend was observed for solvent-sonication extractions (Figure 3.7). A likely explanation of the phenomenon could be that the more arsenic the plant uptakes, the more interactions occur between arsenic and the plant matrix. This may allow more arsenic to be chemically bound to the plant matrix and/or trapped in the plant vacuoles.

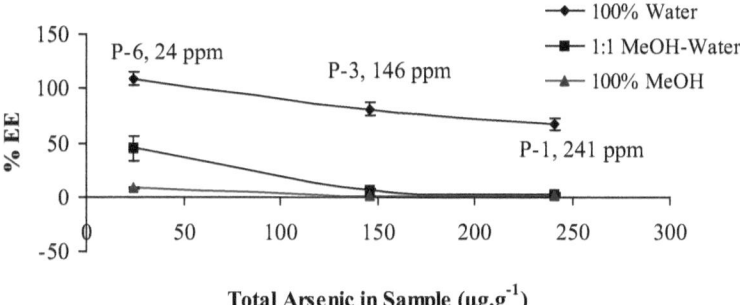

Published in Canadian J. Anal. Spectroscopy, 2002, 47(4), 109.

Figure 3.6. Extraction efficiency as function of total arsenic in plant matrices by Soxhlet-solvent process is shown. Error bars represent ± 1 s.d. from three independent determinations.

It has been demonstrated that both As(III) and As(V) efficiently induced the biosynthesis of phytochelatins ([γ-glutamate-cysteine]$_n$-glycine) synthesized from reduced glutathione (GSH) in plants,[128,129,130] and that arsenic-phytochelatin (As-PC) was present in weak acid extracts.[129,131] As(III) is found to coordinate with sulfur (S) of the thiol groups (SH) of PC in various proportions. As(III)-Glu$_3$ complexes have been identified. The higher arsenic extraction efficiency of dilute HCl might be due its ability to break up the As-S bond of the As-PC complexes. Interestingly, much less arsenic was complexed to phytochelatin in an arsenic hyperaccumulator plant, which is presumably more tolerant to arsenic.[85]

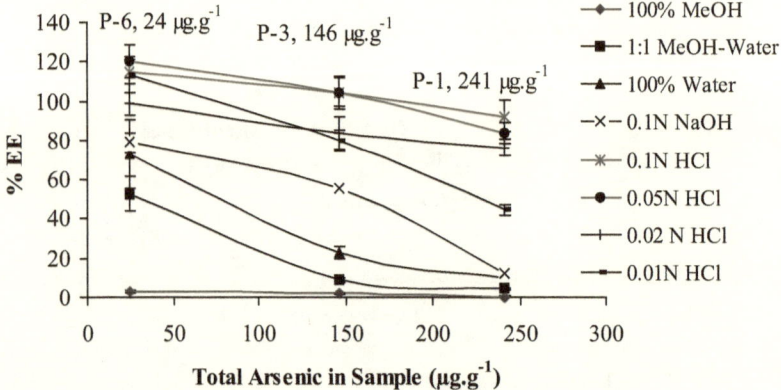

Figure 3.7. Extraction efficiencies as function of total arsenic in plant matrices by solvent sonication process. Error bars represent ± 1 s.d. from 3 independent determinations.

The findings suggested that As formed complexes with the phytochelatins and eventually became sequestered in the plant body. It was also observed that alkaline buffer solutions destabilized As-PC complexes but weak acids could stabilize them.[85] These observations require further study for better understanding of the modes of extraction of As from plant matrices. In Chapter 4 and Chapter 5, more investigations are described for individual plant species and EE versus Concentration relationship.

3.5. MASS BALANCE OF EXTRACTED AND UNEXTRACTED SAMPLES

Arsenic extractions involve a number of steps, and any unaccounted loss from the sample or gain from external sources needs to be examined carefully. Mass balance experiments were conducted by analyzing residues of the water-Soxhlet extract of P-1, P-3 and P-6. Total arsenic in the residues was determined by ashing and digesting with 1:3 v/v HNO_3/HCl acids. Total arsenic of the residue was added to the total of the extract to determine the percent recovery.

Arsenic in Plants: Extraction and Speciation

For the Deloro plants (P-1, P-3 and P-6), the results of arsenic in the original samples, extracts and residues are reported in Table 3.4. Combined arsenic from the extracts and the residues showed quantitative recovery of the analytes from both media. Mass balance experiments were conducted also for the Yellowknife plants and the results are presented in Table 3.5. Arsenic in 1:1 water-methanol and 0.1 M-HCl extracts and arsenic in the residues accounted for all arsenic in the plants quantitatively. Both Deloro and Yellowknife results indicate no loss of arsenic in the analytical process; the arsenic that was not in the extract was in the residue.

Table 3.4. Mass balance experiment performed with water-Soxhlet extracts and residues of plants. Results are mean ± 1 s.d. from 3 independent determinations.

Plant Sample	Total As in Sample (µg)	As in Extract (µg)	As in Residue (µg)	As in Extract and Residue (µg)	% As Accounted
P-1	128 ± 5	85.3 ± 3.9	41.6 ± 3.4	127 ± 7	99 ± 7
P-3	78.5 ± 2.7	63.4 ± 2.4	18.1 ± 2.1	81.5 ± 4.5	104 ± 7
P-6	12.3 ± 0.3	13.4 ± 0.4	0.34 ± 0.16	13.7 ± 0.5	111 ± 4

Table 3.5. Mass balance experiment for Yellowknife plants. Arsenic concentrations in the extracts and residues were determined by ICP-AES. Concentrations are in $\mu g \cdot g^{-1}$.

Plant ID	Total As	A	B	C	D	E	% As Accounted
YK-1	95.2	8.2	20.2	28.4	73.5	102	107
YK-3	18.5	1.8	4.8	6.6	14.6	21.2	115
YK-5	38.5	5.8	20.6	26.4	18.7	45.1	117
YK-7	12.9	0.6	2.9	3.5	10.8	14.3	111
YK-8	4.4	1.2	0.0	1.2	3.1	4.3	98
YK-11	74.8	3.0	19.2	22.2	52.6	75.8	99

A: As extracted by 1:1 Water-MeOH
B: As by 0.05 M-HCl from residue of 1:1 Water-MeOH
C: Sum of As extracted by sequential extraction (C = A+B)
D: As in the residue of sequential extraction
E: Sum of As in extracts and residue (E = C+D)

3.6. EFFECT OF FILTRATION ON ARSENIC CONCENTRATION

Arsenic in the extracts was determined after filtration of the extracts before analysis. The filtration is a necessary step to remove undissolved materials from the liquid samples. In some cases, the filtration process left some residue in the filter paper. Experiments using the Soxhlet extracts of water, 1:1 water-methanol and methanol of P-6 were conducted to determine if any significant difference was caused by the filtration process in the arsenic analysis of the extracts. The arsenic concentrations in the unfiltered acid digested and filtered undigested extracts were determined by HG-AAS. The results are reported in Table 3.6.

The average arsenic concentrations in the unfiltered acid digested and filtered undigested extracts of water-Soxhlet and 50% water-methanol Soxhlet were found to be similar. This shows that for the extracts having comparatively higher arsenic concentrations, filtrations do not contribute significant error in the results. However, for the 100% methanol-Soxhlet extract having much lower arsenic concentrations than the other two extracts, significantly lower arsenic was determined in the filtered undigested extracts indicating loss of arsenic due to filtration.

Table 3.6. Arsenic in unfiltered acid digested and filtered undigested extracts of P-6. Results are mean ± 1 s.d. from 3 independent determinations.

Extracts From	Unfiltered Acid Digested, $\mu g.g^{-1}$	Filtered Undigested, $\mu g.g^{-1}$	t stat (paired t test) ($t_{0.05,2} = 4.30$)
Water-Soxhlet	28.1 ± 3.7	28.8 ± 4.4	0.19
50% MeOH-Soxhlet	15.0 ± 3.0	12.0 ± 2.4	3.59
100% MeOH-Soxhlet	3.47 ± 0.22	2.11 ± 0.08	14.3

Pure methanol was not used in the subsequent extractions but further study should be carried out if 100% methanol is to be used. The problem of filtration with 100% methanol arises mainly because of the dissolved plant organic materials that are often impermeable through the filter paper. Also, methanol's very limited capability of extracting arsenic from plant matrix contributes to the significance of error in the analysis. The residual content of 100% methanol extracts, however, may vary with the types of plants extracted and extraction methods used.

3.7. SPIKE RECOVERY AND SPECIATION

The arsenic concentrations in the plants grown on highly impacted soils are generally high and there are no similar terrestrial plant reference materials with comparable arsenic values. Because there is no agreement on extraction methods, there are no reference values for individual arsenic species in any of the available terrestrial plant reference materials. Marine plants are available but their speciation is vastly different. The predominant arsenic species found in terrestrial plants are As(III), As(V), MA and DMA and therefore these four compounds were used as spikes in a series of experiments designed to assess the effect of solvents and extraction processes on speciation.

Six plant samples were collected in the first phase of sampling from the arsenic impacted sites at Deloro. The three higher arsenic containing plants (P-1, P-3 and P-6) were selected as test samples to conduct extraction experiments. A suitable plant matrix was sought to conduct the spike recovery experiments. Lower arsenic containing plants (P-2, P-4 and P-5) were extracted using 0.05 M-HCl as an extraction medium to determine their potential contribution to the arsenic recovery experiments. The results are given in Table 3.7. Of the three plants, P-5 had the least amount of arsenic (1.83 ± 0.14 $\mu g \cdot g^{-1}$) and 80% of that was extracted. The calculated contribution from the matrix blank to the concentrations of spiked arsenic species was determined to be negligible.

Table 3.7. Amount of arsenic extracted from low arsenic containing plant samples determined by HG-AAS. Results are average ± 1 s.d. from 3 independent determinations.

Plant Sample ID[a]	Arsenic extracted, ($\mu g \cdot g^{-1}$)
P-2	14.6 ± 0.1
P-4	2.61 ± 0.11
P-5	1.47 ± 0.04

[a]Common and scientific names of the plants and their total arsenic content are listed in Table 3.1.

The spike concentrations ranged between 0.20-2.0 $\mu g.g^{-1}$ in solution and were equivalent to concentration of 12-120 $\mu g.g^{-1}$ in the plants assuming complete extraction of 0.5 g of plant samples in 30 mL extractants. Spikes were extracted following the same procedure as that used for the extraction of plant samples. Soxhlet extractions with water and sonication extractions with water, 1:1 water-methanol and 0.05 M-HCl as extractants were carried out.

The concentrations of the inorganic As(III) and As(V) spikes were determined by HGAAS, and the organic DMA and MA spikes by ICPAES. The results of the spike experiments are reported in the Table 3.8. The quantitative recoveries from 90% to 110% were obtained by the extraction media and methods employed. Spike concentrations were also determined by speciation with HPLC-HGAAS method and results are reported in the same table (Table 3.8).

Table 3.8. Results of spike recovery experiment (average ± 1 s.d. from 3 independent determinations). Inorganic and organic spikes were determined by HG-AAS and ICP-AES, respectively.

Extraction medium-method	Spikes	Spikes added, $\mu g.mL^{-1}$	Recovery by ICP-AES, $\mu g.mL^{-1}$	Spike-recovery by HPLC-HGAAS, $\mu g.mL^{-1}$			
				As(III)	As(V)	DMA[1]	MA
1:1 Water-methanol-sonication	As(III)	2.0	2.2 ± 0.1	2.4 ± 0.1	nd		
	As(V)	2.0	2.0 ± 0.2	nd	2.2 ± 0.1		
	DMA	1.0	0.98 ± 0.06			0.94 ± 0.11	
	MA	0.20	0.19 ± 0.02				0.24 ± 0.01
0.05 M-HCl-sonication	As(III)	2.0	2.2 ± 0.1	2.4 ± 0.1	nd		
	As(V)	2.0	2.1 ± 0.1	0.13 ± 0.03	2.4 ± 0.1		
	DMA	0.20	0.195 ± 0.004			0.22 ± 0.01	
	MA	0.20	0.199 ± 0.002				0.27 ± 0.01
100% Water-sonication	As(III)	2.0	2.1 ± 0.1	2.4 ± 0.1	nd		
	As(V)	2.0	2.15 ± 0.04	nd	2.2 ± 0.1		
	DMA	0.20	0.20 ± 0.01			0.21 ± 0.01	
	MA	0.20	0.204 ± 0.004				0.271 ± 0.004
100% Water-soxhlet	As(III)	2.0	2.00 ± 0.01	2.29 ± 0.02	0.06 ± 0.03		
	As(V)	2.0	2.0 ± 0.1	nd	1.8 ± 0.1		
	DMA	1.0	0.88 ± 0.04			1.0 ± 0.1	
	MA	0.10	0.109 ± 0.001				0.13 ± 0.01

Void spaces and 'nd' represent species not detected. [1]DMA results were obtained after optimization of DMA analysis (Section 3.8).

Arsenic in Plants: Extraction and Speciation

Overall data show good recovery and that the extraction procedures used are mild and do not significantly alter the arsenic speciation. Oxidized or reduced forms of the introduced inorganic species were not detected in the 1:1 water-methanol sonication and 100% water-sonication extracts. However, in the 0.05 M-HCl-sonication and 100% water Soxhlet extracts a small amount of interconversion of the inorganic arsenic species was observed.

Topics on the stability of arsenic species particularly the interconversion of As(III) and As(V) have received detailed treatments in the reviews.[6,27] However, in plant extracts - a broth containing a plethora of metal ions and plant organic materials - identification of a single or a few factors responsible for the arsenate to arsenite ratio may be difficult. While bacteria, O_2, H_2S, and metal ions such as Fe^{3+}/Fe^{2+}, play definite roles in the conversion of arsenic species in aqueous media, temperature, time and storage conditions have also been implicated.[6]

A great deal of uncertainty in the stability of As(III) has been observed and the changed product has been almost certainly the As(V).[27] These observations have importance with respect to the health risks associated with arsenic. The stability or instability of the As(III)/As(V) pair does not make any crucial change in terms of its toxicity; the transformation only switches between the two highly toxic forms.

Our study, however, showed that the species were stable throughout the extraction and speciation for the methods and techniques employed (Table 3.8). The standard stocks in 10 mg.L^{-1} or higher concentrations in plastic bottles stayed stable for months under normal (5-6°C) refrigeration conditions. We observed good qualitative correlation between the species found in the original plant samples and the extraction residues by x-ray absorption analysis (XANES, discussed in Ch. 2 and Ch. 5), and those found in the extracts by HPLC-HGAAS.

3.7.1. Recovery of Mixed Spikes

The spike recovery experiments were previously conducted using the standard arsenic spikes independently. In nature, however, arsenic species coexist in plant matrices. Mixed spikes recovery analysis was, therefore, carried out by using standard concentration 0.20 mg.L^{-1} each of the four common arsenic species, As(III), As(V), MA and DMA spiked on a grass matrix. The grass sample was collected from a low arsenic background location and prepared by washing, drying and grinding similarly to the other plant samples. The experiment was conducted to investigate the recovery pattern of the arsenic species by the sequential extraction method from the plant matrix. The standard spikes on grass matrix were speciated by HPLC-HGAAS and results are reported in Table 4.10 along with the results of plant sample analyzed at the same time. The concentrations of the recovered species were determined to be 0.18, 0.15, 0.21 and 0.19 mg.L^{-1} for As(III), DMA, MA, and As(V), respectively. The average recovery was more than 91% in 1:1 water-methanol medium while ICPAES analysis in the 0.1 M-HCl extract did not detect any arsenic.

The quantitative extraction of both organic and inorganic arsenic in the 1:1 water-methanol medium suggested that there was little or no chemical and/or physical attraction of the spiked species for the plant matrix. On the other hand, arsenic species in natural samples cannot be completely extracted in one solvent showing their considerable physical and/or chemical association with the matrix.

3.8. DMA EXPERIMENT

The amount of DMA used in the spiked experiments was determined by ICPAES and showed 100% (quantitative) recovery; however, this was not confirmed quantitatively by HPLC-HGAAS. Consistently, much higher values than the introduced amounts were observed. Experiments were conducted to assess the reason for the high DMA results in the HPLC-HGAAS determinations. For this, calibration standards of DMA were prepared as follows: DMA in DDW, DMA with MA in DDW, and DMA with MA, As(III) and As(V) in DDW. Independent calibration plots were prepared for DMA in the mixtures by HPLC-HGAAS runs and are shown in Figure 3.8. The slopes of the calibration curves obtained from the DMA with MA, and DMA with MA, As(III) and As(v) were identical.

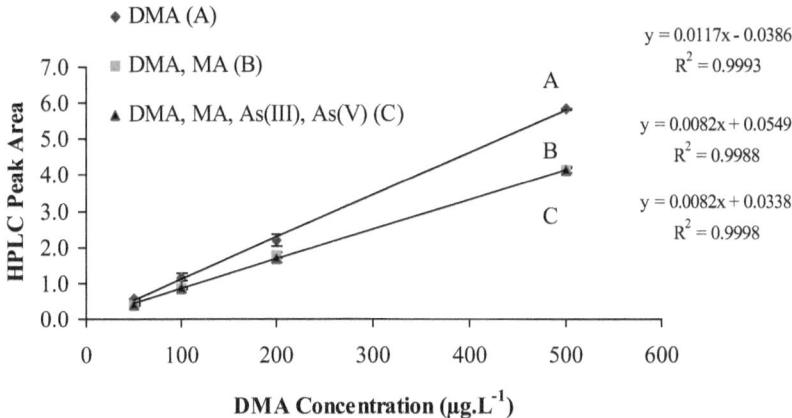

Figure 3.8. DMA calibration curves prepared with different arsenic calibration standards are presented. Error bars represent ± 1 s.d. from 3 independent determinations.

On the other hand, the slope of the curve obtained from independent DMA calibration was higher. DMA concentrations calculated using the linear equation obtained from the latter curve were in good agreement with the introduced DMA concentrations. The other curves yielded much higher concentrations than the introduced concentrations of DMA due to smaller sensitivity (slope) of the calibration line. Therefore, the spiked DMA concentrations, as shown in the Table 3.8, were calculated using the independent DMA calibration curve. It should be noted that the concentrations of the other arsenic standards were not affected by the mixed standard calibration methods.

Arsenic standards were prepared in DDW using appropriate amounts from the concentrated commercial stocks. It was not fully known how the plant matrix would have influenced the arsenic speciation when standards were prepared in DDW for calibration. Though it may appear impractical in the speciation of arsenic in many plant samples, it would be worthwhile investigating the matrix effect on arsenic speciation using a limited number of plants. For the recovery experiments of DMA spiked on P-5 plant matrix, preparation of DMA standards in blank P-5 extract solution improved results of DMA significantly. For example, in the same HPLC-HGAAS experiment, 300 µg.L^{-1} DMA prepared in DDW was determined to be 155 µg.L^{-1}, almost one-half the original concentration. Conversely, DMA concentrations spiked on P-5 for different extraction methods (Table 3.8) showed quantitative (100%) recovery calculated using linear plots obtained from the DMA calibration standards prepared in blank P-5 water-Soxhlet extracts.

3.9. SPECIATION OF ARSENIC IN PLANT EXTRACTS

The extracts from P-1, P-3 and P-6 were analyzed for arsenic species by HPLC-HGAAS. The concentrations of total arsenic in the extracts and individual species are shown in Table 3.9. The total concentrations of species determined by HPLC-HGAAS agree well with the total arsenic in the extracts determined by ICP-AES or HG-AAS.

Table 3.9. Speciation of arsenic extracted from plant samples. Results are average ± 1 s.d. from 3 independent determinations. Concentrations are in μg.g^{-1}.

Extraction method	Plant ID	Amount of As Extracted	As by HPLC-HGAAS[a] As(III)	As(V)	Sum of species
100% water-sonication	P-1	24 ± 2	nd	21 ± 1	21 ± 1
	P-3	34 ± 3	nd	29 ± 3	29 ± 3
	P-6	18 ± 4	13 ± 1	6.3 ± 0.2	19 ± 1
0.01 M-HCl-sonication	P-1	108 ± 2	5.9 ± 1.1	96.1 ± 0.2	102 ± 1
	P-3	117 ± 4	11.9 ± 0.9	97 ± 5	109 ± 9
	P-6	28 ± 2	16.7 ± 0.1	10 ± 1	27 ± 1
0.02 M-HCl-sonication	P-1	184 ± 3	24 ± 14	169 ± 24	193 ± 38
	P-3	122 ± 8	36 ± 3	100 ± 21	136 ± 24
	P-6	24 ± 1	15 ± 1	12.3 ± 0.2	27 ± 1
0.05 M-HCl-sonication	P-1	202 ± 6	15 ± 1	184 ± 6	199 ± 7
	P-3	153 ± 6	29 ± 1	114 ± 6	143 ± 7
	P-6	29.3 ± 1.3	21.6 ± 0.3	12.5 ± 1.4	34 ± 2
0.1 M-HCl-sonication	P-1	223 ± 11	12 ± 1	208 ± 8	220 ± 9
	P-3	153 ± 7	31 ± 8	143 ± 8	174 ± 16
	P-6	28 ± 1	8 ± 2	27 ± 1	35 ± 3
100% water-soxhlet	P-1	160 ± 7	nd	150 ± 16	150 ± 16
	P-3	118 ± 4	nd	100 ± 14	100 ± 14
	P-6	26 ± 1	nd	22 ± 4	22 ± 4
1:1 water-methanol-sonication	P-1	7.7 ± 1.0	1.7 ± 0.8	5.6 ± 0.4	8.8 ± 1.2[b]
	P-3	10.0 ± 0.5	5.6 ± 1.5	6.5 ± 0.8	12 ± 2
	P-6	9.7 ± 0.5	10.8 ± 0.6	2.1 ± 0.3	13 ± 1

[a]MA was detected only in the water-methanol extracts of P-1 (1.51 ± 0.04). [b]Sum includes MA concentration. DMA was not detected in any extract, and nd = not detected.

The majority of the arsenic extracted was inorganic. However, MA was detected in P-1 extracted by 1:1 water-methanol solvent only (Table 3.9). Though the EE of 1:1 water-methanol was low, the only organoarsenic species detected in the target plants was extracted in this medium.

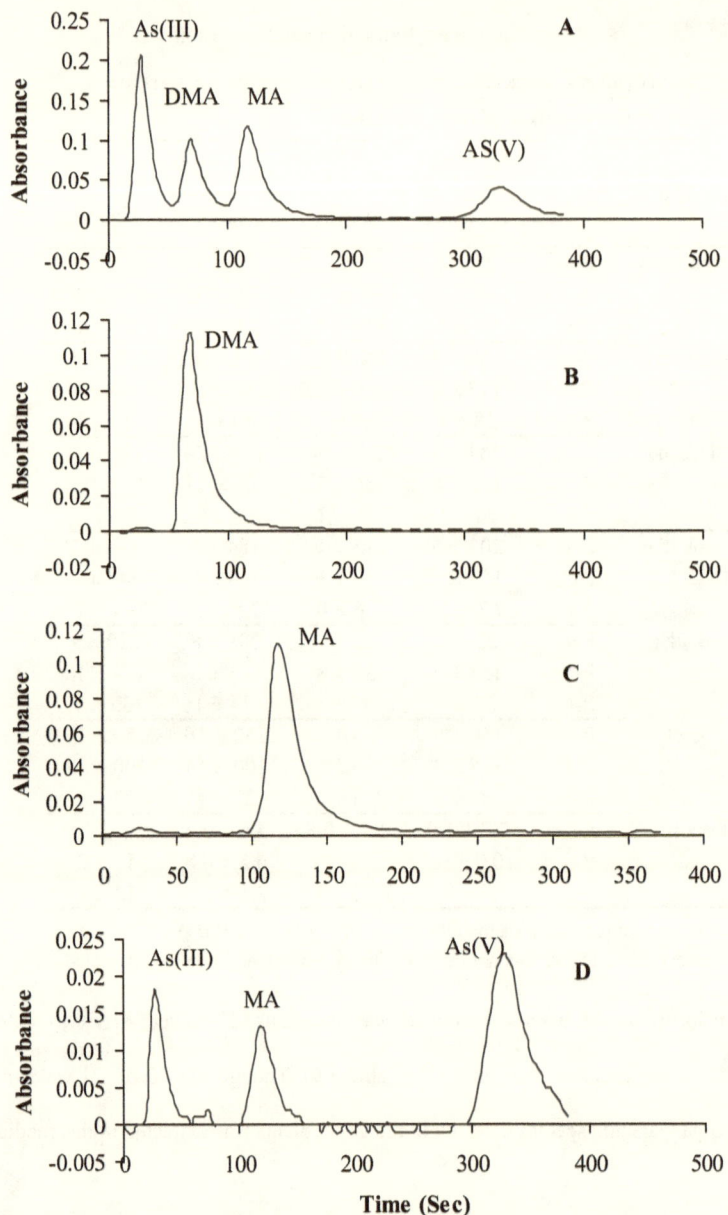

Figure 3.9. Representative HPLC-HGAAS chromatograms. A: Standard arsenic species; B: DMA spike; C: MA spike; and D: 1:1 water-methanol extract of plant P-1 (field horsetail). HPLC was performed with anion exchange column (Hamilton PRP-X100 250 x 4.6 mm column), 20 mM ammonium phosphate, pH 6.0 at 1.5 mL.min^{-1}.

Arsenic in Plants: Extraction and Speciation

This observation was important since it indicated that the low concentration or trace organoarsenic species could be detected in alcohol media where inorganic species are strongly suppressed from extraction from plant matrices. The representative chromatograms of standard arsenic species, spikes MA and DMA along with the chromatogram of P-1 extracted in 1:1 water-methanol are shown in Figure 3.9. In terrestrial plants, organoarsenic species that are usually encountered are MA and DMA.[132] Up to 2% MA and DMA were measured in shoots and roots of *Holcus lanatus* and *Arabidopsis thaliana*.[133]

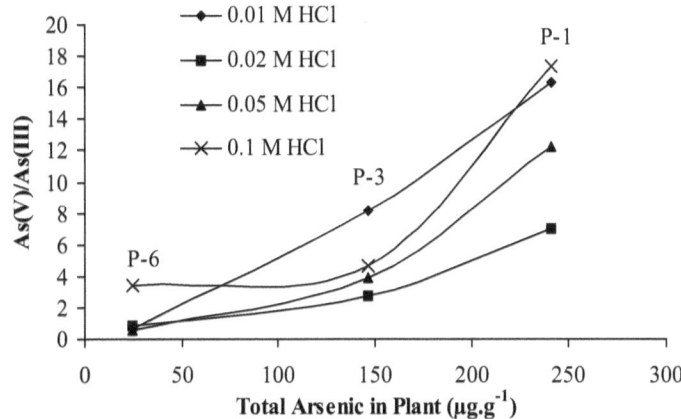

Figure 3.10. Ratio of arsenate to arsenite extracted in dilute HCl solutions as function of total arsenic in the plants.

In the acid media, concentrations of As(V) predominated over that of As(III) in all cases except for plant P-6. For P-6, As(III) predominated over As(V) in all but the 0.1M-HCl medium. No As(III) was detected in water-Soxhlet extracts of all plants indicating the oxidizing environment/capability of Soxhlet extraction. The reducing and/or stabilizing nature of chloride ions for As(III) was shown by the presence of As(III) in all HCl media. Both As(III) and As(V), were also present in 1:1 water-methanol extracts of all plants indicating the milder nature of the water-alcohol solvent.

From the results of the current experiments (Table 3.9), however, a general trend of increase in the arsenate to arsenite ratio with total arsenic content in the extracts of acid media was observed, Figure 3.10. An explanation for the increase in the As(V) fraction may be found from the XANES results of plants discussed in Chapter 5.

3.10. LIMITS OF DETECTION AND PRECONCENTRATION OF ARSENIC SPECIES

A reliable detection limit (RDL) of the standard arsenic species in solution for HPLC-HGAAS was determined by running 50 $\mu g.L^{-1}$ standard samples eight times in a row and calculating the standard deviations of the concentrations of the species. The RDL of each species was determined by using the formula, RDL = 2 x (t-stat) x (stdev.). The RDL results are reported in Table 3.10. The RDL's of As(III), As(V), MA and DMA were 7, 31, 15 and 14 $\mu g.L^{-1}$, respectively.

Table 3.10. Reliable detection limits (RDL) of arsenic species determined by HPLC-HGAAS. RDL = 2(t-stat)(s.d.).

Arsenic Species	RDL ($\mu g.L^{-1}$)
As(III)	7
As(V)	31
MA	15
DMA	14

The method detection limits (MDL) of ICPAES and HGAAS were determined by calculating the standard deviations of the lowest standard concentrations, 0.10 $mg.L^{-1}$ and 5.0 $\mu g.L^{-1}$ for ICPAES and HGAAS, respectively, measured by the instruments over the period. The results are presented in Table 3.11. The MDL was defined by the formula, MDL = t-stat x (standard deviation) and, for ICPAES and HGAAS were determined to be 0.0148 $mg.L^{-1}$ and 0.186 $\mu g.L^{-1}$, respectively.

Table 3.11. Determination of method detection limits (MDL) of ICP-AES and HG-AAS. MDL = (t-stat)(s.d.).

ICP-AES		HG-AAS	
Date of Expt.	Concentration, mg.L^{-1}	Date of Expt.	Concentration, µg.L^{-1}
6/12/2003	0.1199	1/16/2002	4.551
7/10/2003	0.1064	3/27/2002	4.519
7/16/2003	0.1224	4/12/2002	4.313
8/13/2003	0.1043	4/15/2002	4.463
9/9/2003	0.1069	4/25/2002	4.411
9/14/2003	0.1071	7/1/2002	4.398
10/1/2003	0.0984	10/2/2002	4.266
11/28/2003	0.1137	6/21/2004	4.476
12/9/2003	0.0950		
12/12/2003	0.0965		
1/21/2004	0.1002		
2/25/2004	0.1055		
8/8/2004	0.0970		
8/10/2004	0.1040		
MDL (95% CL), mg.L^{-1}		MDL (95% CL), µg.L^{-1}	
0.0148		0.186	

The organoarsenic compounds, MA and DMA, were very minor components of the total arsenic determined in most of the plants. Their concentrations in original 30 mL extracts would be much below the detection limits of HPLC-HGAAS method for the species. Since 1:1 water-methanol extracts were dried completely to remove all solvents, the dried extractants could be regenerated in a much smaller volume of DDW. For example, the dissolution of dried extracts of original 30 mL extracts in 3 mL DDW would result in a concentration factor of ten. This process enabled reliable detection of MA and DMA in many plant samples by HPLC-HGAAS.

3.11. THE WATER CONTENT OF PLANTS

In order to determine the accurate analyte concentration the correct water content of samples must be known. All plants do not contain water at the same amount and may not dry at the same rate. In order to drive all water out of the sample matrix, oven drying above 100°C is necessary though may not always be practical. The experiment was carried out using the Yellowknife plants by drying them in the open air and then in an oven at 105°C. The samples were weighed from time to time and returned for further drying. A plot of percent water loss against time of drying is given in Figure 3.11.

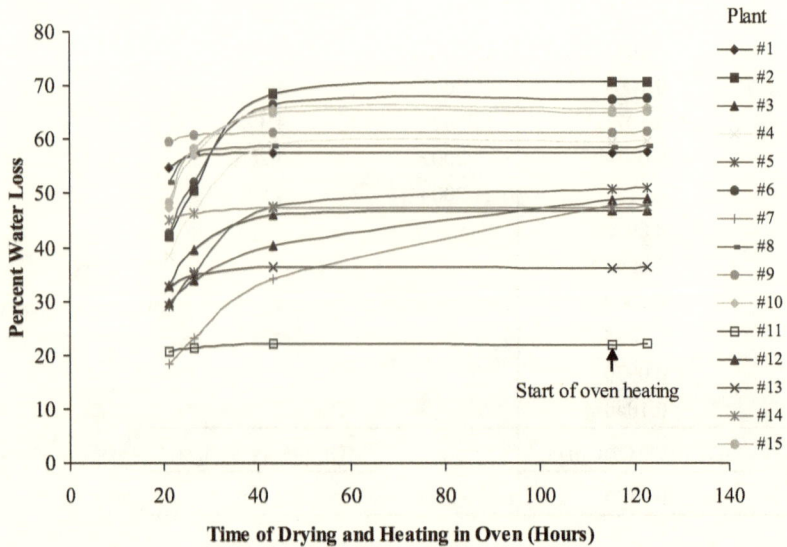

Figure 3.11. Determination of loss of water from plant matrices as function of time of drying in open air and heating in oven at 105°C. Fifteen Yellowknife plant samples were dried for this experiment. The plants are listed in details in Table 3.26.

The water content of plants ranged from about 20% to about 70%. When no more significant loss of weight was observed after air drying, the drying was considered complete. Only a few of the plants completed drying after 20 hours and many plants lost only about half of their total water content. Most of the plants completed drying after 40 hours and only a few apparently continued drying after 80-100 hours. To determine whether all water was removed after about 120 hours of air drying, the plants were placed in an oven at 105°C for further drying under the heat. There was not any significant loss of weight due to heating at 105°C indicating that the open-air drying method was adequate for removing water from plants. It should be noted also that the rate of drying in the open air would depend on the particle size of sample matrix, the shape and size of the drying container as well as on the humidity of the room air.

After washing, plant samples collected from Deloro were dried in the open air for 3 or 4 days initially and then dried again in the oven at a moderate temperature of about 70°C. This process made the plant samples crisp and easy to grind. The humidity of laboratory air and the nature of plant matrix had a significant influence on the drying time of plant samples in the open-air drying process. Although the bench top open-air drying is a practical alternative to oven drying of many plant samples, there is no alternative to knowing the correct water content for the determination of accurate analyte concentrations. Since it is not possible to control laboratory humidity at a constant level all the time, also since all fresh plants do not hold the same amount of water or dry at the same rate, equal time of air drying may not lead to equal amount of drying for all plants. Therefore, open air drying need to be monitored carefully. For reporting very accurate analyte concentrations, however, oven drying of a known amount of sample is necessary.

3.12. ADSORBED OR ABSORBED ARSENIC IN PLANT MATRIX

The rationalization for distinguishing the absorbed and adsorbed toxic contaminants in plant matrices is not important when consumption of plants is considered; however, it is an important distinction to understand mechanisms of uptake and storage of arsenic in the plant. Wildlife does not consume 'washed' plants in the natural environment. On the other hand, the criterion for washing to preserve analyte toxicants from being washed out of the plant matrix is not well defined. Washing samples of plant origin has been reported to produce variable results. Washing techniques with various solvents including 0.1 M-HCl and 0.02% detergent have been evaluated for removal of external fluoride from ironbark and grape leaves. There was not any significant difference between hand washing and bag washing of the samples and no leaching of the analyte has occurred due to washing.[134] Washing of visually clean pecan leaves did not improve the analysis of trace nutrient; rather contamination from wash-water was observed.[135]

However, differences between washed and unwashed plant tissue samples have been observed with the unwashed samples showing minor contamination which was readily removed by washing with clean running water.[136]

Even after washing with copious amounts of water, it was not known if all adsorbed arsenic was removed or not. In order to verify, a part of the samples after washing with normal washing procedures were further washed with 0.1 M-HCl. The results of total arsenic from the water washed, acid rinsed samples were compared, and any significant difference between the results was not observed. A field horsetail (DL-14) was further washed with 0.1M-HCl after regular washing with water and analyzed for total arsenic. While average (n = 3) total arsenic of 299 ± 20 $\mu g.g^{-1}$ was determined in the water washed samples, a content of 286 $\mu g.g^{-1}$ arsenic was found in the acid rinsed sample with a loss of 4% of the original amount and within the error limits of the original average arsenic content. Considering the high EE of HCl from the field horsetail (see Figure 3.3), the loss should be considered insignificant. This result showed that careful washing of the plants with copious amounts of water would be sufficient to remove the adherent soils and contaminants from the plants. There should be minimum cuts and bruises before and during washing to minimize the loss of internal materials from plant samples.

3.13. EFFECT OF SOAKING ON EXTRACTION EFFICIENCY (EE)

Plant samples tend to absorb liquid solvents and swell up, and in the process may open up plant tissues and cell vacuoles to the extractants for the analytes. In order to assess whether soaking would improve EE, plant samples were mixed by vortexing in 10 mL extractant and left to soak overnight for 20 hours in solvents before extraction. Three solvents, 0.1 M-HCl, 1:1 water-methanol and Gastric fluid, were tested on four plant samples (Table 3.12).

Table 3.12. Determination of the effect of soaking on extractability of arsenic. All concentrations are in $\mu g.g^{-1}$. Percent extraction efficiencies are given in parentheses.

Plant ID	0.1 M-HCl		1:1 H$_2$O/MeOH		Gastric fluid
	20 Hours Soaked	Normal	20 Hours Soaked	Normal	20 Hours Soaked
YK-1	31.2 (33)	26.6 (28)	7.7 (8)	8.18 (9)	29.7 (31)
YK-3	7.6 (41)	6.7 (36)	2.8 (15)	1.81 (10)	4.4 (24)
YK-5	23.9 (62)	26.4 (69)	5.7 (15)	5.8 (15)	23.4 (61)
YK-11	28.7 (38)	22.2 (29)	3.7 (5)	3.38 (4)	24.5 (32)

The amount of arsenic extracted and the %EE in parentheses are given in the Table. Results from normal extractions (i.e., without soaking overnight) for acid and water-methanol solvents are also presented. A comparison between the normal and soaked extraction results of the plants showed no significant improvement due to soaking. Further experiments with the enzyme (pepsin) and their results are presented in Section 3.16.

3.14. EXTRACTION WITH ORTHOPHOSPHORIC ACID

The extraction and speciation of arsenic by microwave assisted 0.3 M-H$_3$PO$_4$ and HPLC-HGAFS from terrestrial plants (such as tobacco leaves, rice, bean and hot pepper, etc.) has been reported.[83] In the study, the capability of phosphoric acid to extract both organic and inorganic arsenic species was demonstrated. In a quest to see if phosphoric acid aided by sonication could be used for extraction and speciation by HPLC-HGAAS, a number of plants from Yellowknife were extracted with 0.3 M-H$_3$PO$_4$. Extracts were analyzed for total arsenic by ICPAES. An aliquot of the extract (10 mL) was reduced to 2 mL by evaporation (de-watered extracts) to improve the concentration factor of the trace organoarsenic species present. Arsenic speciation by 0.3 M-H$_3$PO$_4$ and concentrated extracts was carried out by HPLC-HGAAS. These results are reported in Table 3.13 along with sequential (1:1 water-methanol and 0.1 M-HCl) and 1.0 M-HCl speciation results for comparison. The extraction efficiencies of 0.3 M-H_3PO_4 sonication extraction were comparable with those of the sequential method from ICPAES determinations as shown in Table 3.14.

From the speciation results, it can be seen, however, that very little As(III) was detected in 0.3 M-H_3PO_4 extracts and no As(V) in 0.3 M-H_3PO_4 de-watered extracts. Unlike 1:1 water-methanol, de-watered 0.3 M-H_3PO_4 extracts experienced severe difficulty in the speciation of arsenic in the plants.

Table 3.13. Comparison of arsenic extraction and speciation by 0.3 M H_3PO_4 with those of sequential and 1.0 M-HCl. Concentrations are in $\mu g \cdot g^{-1}$.

Sample ID	As(III)				MA			As(V)			
	Sequential Method	1.0 M-HCl	0.3 M-H_3PO_4	0.3 M H_3PO_4 (dw)[a]	0.3 M-H_3PO_4	0.3 M-H_3PO_4 (dw)[a]	1:1 Water-MeOH	Sequential Method	1.0 M-HCl	0.3 M-H_3PO_4	0.3 M-H_3PO_4 (dw)[a]
YK-1	3.1	2.7 ± 0.1[b]	nd	1.36	0.48	nd	0.36	23.8	13.0 ± 0.1[b]	22.3	nd
YK-4	0.53	0.1	nd	0.30	nd	nd	0.23	2.9	0.2	0.58	nd
YK-5	9.1	3.4	0.41	0.15	nd	nd	nd	24.3	10.5	17.8	nd
YK-11	0.80	3.1	nd	2.27	nd	nd	0.26	24.8	12.9	15.6	nd
YK-11 dup.	0.59	2.8	nd	4.50	0.60	nd	0.23	26.8	11.3	17.6	nd

[a]De-watered extracts were regenerated in small vol. of DDW. dw = de-watered, nd = not detected.
[b]Mean ± ave. deviation from 2 independent determinations.

Table 3.14. Comparison of extraction efficiencies of the sequential, 1.0 M-HCl, 0.3 M-H_3PO_4 and 0.3 M-H_3PO_4 (de-watered) extractions determined by ICP-AES and HPLC-HGAAS.

Sample ID (Total As, $\mu g \cdot g^{-1}$)	% EE from ICP-AES Analysis		% EE from HPLC-HGAAS Analysis			
	Sequential Method	0.3 M-H_3PO_4	Sequential Method	1.0 M-HCl	0.3 M-H_3PO_4	0.3 M-H_3PO_4 (de-watered)
YK-1 (95)	30	35	29	16	24	1.4
YK-4 (8)	38	60	45	3	7	4
YK-5 (38)	68	72	87	36	47	0.4
YK-11 (78)	31	43	36	20	21	3

Water-alcohol extracts can be easily dried and regenerated for improved concentration factors for the trace organoarsenic species and for analysis by the HPLC-HGAAS method. The drawbacks of the phosphoric acid method were its inability to detect MA in plant YK-4 at all and its inconsistent results obtained for MA in YK-11. In one sample it showed 0.60 µg.g^{-1} MA but did not detect any in the duplicate sample. On the other hand, the 1:1 water-methanol method showed consistent results of 0.23 and 0.26 µg.g^{-1} in the duplicate samples. Further, the overall species stability of 0.3 M-H_3PO_4 extracts was found to be inferior to those in the water-alcohol and HCl media.

3.15. EXTRACTION BY 100% METHANOL, 50% METHANOL AND WATER

Extraction solvents are generally prepared with water (polarity index 10.2). Water's solvent properties are modified by other reagents (e.g. acid and alcohol) to affect the extraction efficiency of the solvent. Among the protic solvents, methanol (polarity index 5.1) is a close neighbor of water in terms of polarity. These solvents are miscible. Methanol is preferred to ethanol due to its lower viscosity and boiling temperature. Both water and water-methanol mixtures at various proportions have been used for the extraction of arsenic from plant and animal tissues.[15,16,137] In plants grown on arsenic contaminated soils, the concentrations of arsenic are higher than those in the background plants and most of the arsenic is inorganic.

The sequential extraction of arsenic by 1:1 water-methanol and 0.1 M- or 0.05 M-HCl showed that the organoarsenic compounds were detected in the water-alcohol media only. On the other hand, HCl was more efficient in extracting the inorganic arsenic. The organoarsenic species are in much smaller amounts than the inorganic species in most terrestrial plants and their determination becomes difficult in the presence of the large amounts of inorganic arsenic as demonstrated in Figure 3.12. The potential application of water/methanol as extractor/separator of organic species from the inorganic species of arsenic has been investigated.

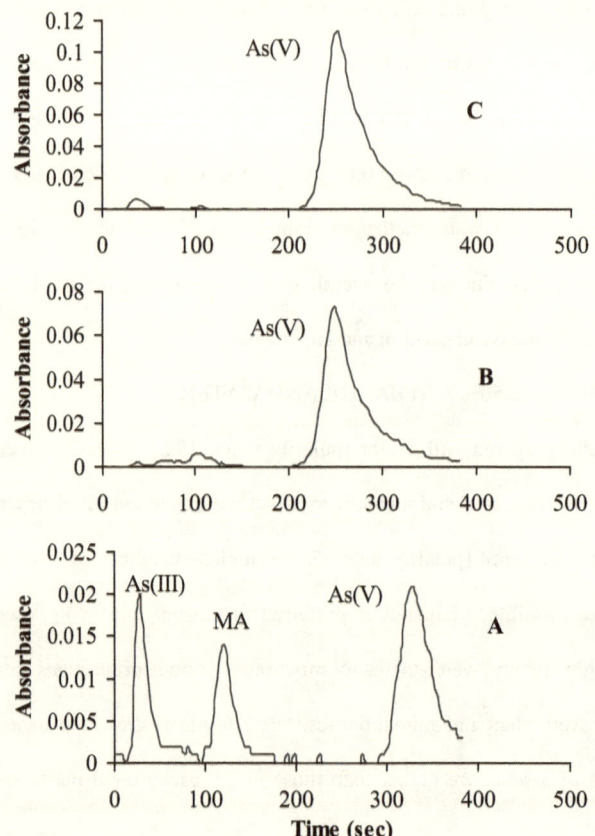

Figure 3.12. HPLC-HGAAS chromatograms of 1:1 water-methanol (A), water (B) and 0.1 M HCl (C) extracts of Field horsetail (P-1). HPLC of water extracts (B), acid extracts (C) were performed in June (same day), and HPLC of 1:1 water-methanol extracts (A) in August of the same year. HPLC conditions are given in Figure 3.9.

A series of experiments was conducted using 100% methanol, 50% methanol (1:1 v/v water-methanol) and 100% water (DDW) for extracting arsenic from a number of plant samples from Yellowknife and Deloro. All extracts were evaporated to dryness in the oven at low temperature (~65°C) and regenerated in 3 mL DDW by shaking overnight on a bench top shaker. Speciation experiments were conducted by HPLC-HGAAS and the results of speciation are reported in Table 3.15.

Arsenic in Plants: Extraction and Speciation

Table 3.15. Results of 100% methanol, 50% methanol and water extracts by HPLC-HGAAS. Results are mean ± ave. deviation from 2 independent determinations. All concentrations are in $\mu g.g^{-1}$ and void spaces represent species not detected.

Sample ID	100% MeOH			50% MeOH (1:1 v/v Water-Methanol)				100% Water	
	As(III)	DMA	MA	As(III)	DMA	MA	As(V)	As(III)	As(V)
YK-1	3.3 ± 0.1					0.20 ± 0.01	4.9 ± 0.3		7.5 ± 0.3
YK-4	0.55 ± 0.04		0.13 ± 0.01	1.48 ± 0.01		0.14 ± 0.04			1.5 ± 0.1
YK-11	0.79 ± 0.12			2.1 ± 0.2			0.24[a]		4.2 ± 0.3
DL-105	1.0 ± 0.3	0.57 ± 0.06				0.45 ± 0.01	2.20 ± 0.04	2.6 ± 0.5	3.2 ± 0.2
DL-129	0.53 ± 0.05			0.60 ± 0.11			1.1 ± 0.1	2.26 ± 0.03	1.17 ± 0.04

[a] As(V) was not detected in the duplicate sample of YK-11 only.

Table 3.16. Total arsenic extracted by the solvents. Results are mean ± ave. deviation from 2 independent determinations. Concentrations are $\mu g.g^{-1}$.

Sample ID	Total from 100% MeOH	Total from 50% MeOH	Total from 100% water
YK-1	3.3 ± 0.1	5.1 ± 0.3	7.5 ± 0.3
YK-4	0.68 ± 0.05	1.6 ± 0.1	1.5 ± 0.1
YK-11	0.79 ± 0.12	2.3 ± 0.2	4.2 ± 0.3
DL-105	1.6 ± 0.4	2.7 ± 0.1	5.8 ± 0.7
DL-129	0.53 ± 0.05	1.7 ± 0.2	3.4 ± 0.1

Important features of the results are the detection of no As(V) in 100% methanol and no organoarsenic species in 100% water extracts. In 50% methanol, all four species of arsenic were detected. The total amounts of arsenic extracted by the solvents increased from 100% methanol to 100% water as reported in Table 3.16. This increase of extracted arsenic correlated with the increase of polarity of the solvents. A 3-D representation of the amount of extracted arsenic versus the polarities of the solvents is shown in Figure 3.13. The polarities of water and methanol were taken from the text book table[138] while that of 50% methanol was taken as the average of polarities of the two pure solvents. The increased amounts of extracted arsenic with increased polarity were observed for the plants. The increased EE may be due to the increase in the polarity of the solvent, with more polar As(V) tending to partition in the more polar medium.

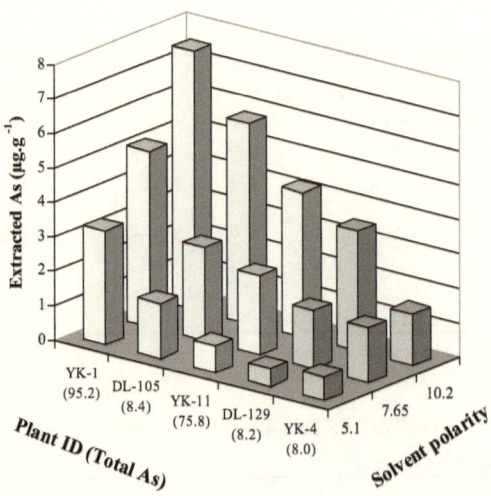

Figure 3.13. The bar graphs show amounts of arsenic extracted as function of polarity of the solvents.

The HPLC-HGAAS chromatograms of 100% methanol, 50% methanol and 100% water extracts Yellowknife and Deloro plants are shown in Figure 3.14. The HPLC-HGAAS chromatograms of duplicate or triplicate replicas of independent extracts of each plant were compared. The nearly identical replication of the chromatograms of each plant

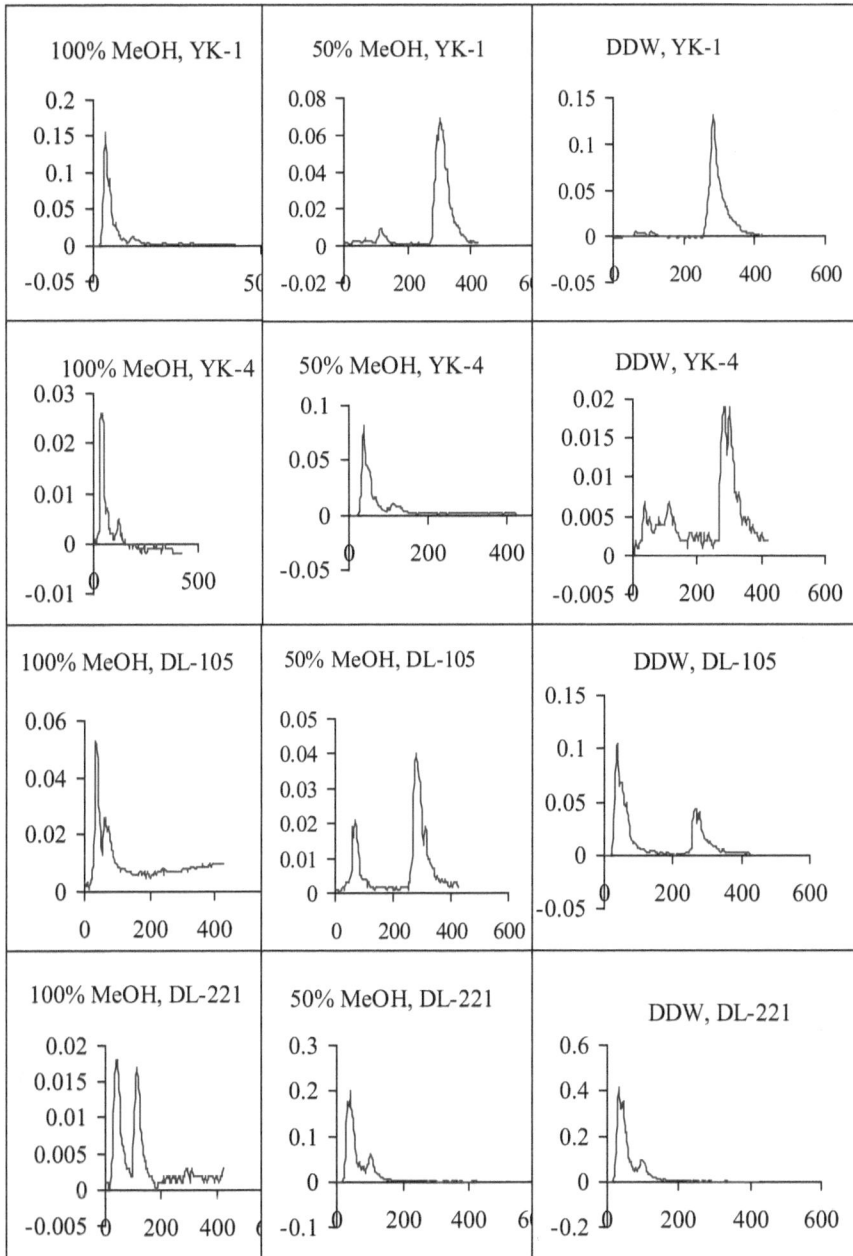

Figure 3.14. HPLC-HGAAS chromatograms (Absorbance vs. Time in Sec.) of Yellowknife and Deloro plants extracted in 100% methanol, 50% methanol and DDW. Nearly identical chromatograms (not shown here) were obtained for duplicate and triplicate independent analyses of the samples indicating stability of the arsenic species throughout the analytical procedure. HPLC conditions are described in Figure 3.9.

reveals that the speciation of arsenic in these solvents was not arbitrary. This showed that the process of extraction and evaporation of the solvents to dryness, the regeneration of the extracts in DDW, and finally, the determination of species with HPLC-HGAAS had little influence on the speciation of arsenic.

The DMA concentrations in ten plants extracted by 100% methanol and 50% methanol (1:1 v/v water-methanol) are given in Table 3.17.

Table 3.17. Extraction efficiencies of 100% MeOH and 50% MeOH for DMA. All but DL-129 show greater DMA in 100% MeOH than in 50% MeOH.

Plant ID	DMA Extracted in		t-test Paired Two Samples for Means		
	100% MeOH, $\mu g \cdot g^{-1}$	50% MeOH, $\mu g \cdot g^{-1}$		100% MeOH	50% MeOH
DL-105	0.506	0.397	Mean	0.242	0.149
DL-108	0.202	0.131	Variance	0.0124	0.0153
DL-119	0.258	0.234	Observations	12	12
119 Dup	0.366	nd	df	11	
DL-129	0.176	0.234	t Stat	2.93	
DL-141	0.248	0.097	t Critical one-tail	1.80	
141 Dup	0.168	0.111			
DL-205	0.203	0.107			
DL-206	0.198	nd			
DL-209	0.286	0.277			
DL-220	0.235	0.206			
DL-221	0.057	nd			

An inspection of the results of extraction of DMA by the solvents reveals that enhanced extraction efficiency was attained by 100% methanol. For some plants, DL-119 duplicate sample and DL-206 for example, DMA was not detected in 50% methanol extracts while in 100% methanol DMA showed significant presence. The Student's t-test for paired samples showed a significant difference between the means of the two extractants. During the speciation, the elution of DMA occurs right after As(III) and before MA on the HPLC column (Hamilton PRPx100) (Figure 3.9A) with the phosphate buffer as elution solvent. This sequence indicates DMA has a lower polarity than MA and is likely due to DMA's structure of two methyl groups as compared to MA's one such methyl group. A reason for the better extraction of DMA in the 100% methanol extracts than in the 50% methanol extracts may be explained by the fact that the polarity of 100% methanol was less than 50% methanol and DMA having two alkyl groups in its structure was partitioning better in this solvent (i.e., 100% methanol).

3.16. GASTRIC FLUID EXTRACTION (GFE)

The bioavailability of arsenic in the human stomach and intestine is dependent on the EE of the gastric fluid in the system as described in Chapter 1. Gastric fluid extractions are important in order to evaluate the risks associated with the consumption of food contaminated by toxic elements, and the development of the in vitro synthetic gastric fluid extractions has been discussed earlier (Section 1.7.2.2).

In this study following a method published earlier,[94] GFE was carried out using synthetic gastric fluid solution consisting 1.25 g.L^{-1} pepsin (Sigma P7000: activity 1:10000) and 8.77 g.L^{-1} NaCl. The pH of the solution was adjusted to 1.8 with concentrated HCl. A 0.5 g (± 0.1 mg) powdered plant sample was weighed into 50 mL centrifuge tube and 20 mL of gastric fluid was added. After placing the lids securely, the samples were vortexed for one minute and then shaken at 250 rpm for 60 minutes at 37°C in an incubator shaker (Innova Refrigerated Incubator Shaker, New Brunswick, Edison, N.J., USA). The resulting solutions were centrifuged at 3000 rpm for 10 minutes and the supernatants were decanted in 100 mL plastic bottles. The extracts were filtered and analyzed for total arsenic and arsenic species.

In this study, the GFE aided by sonication was carried out to ascertain extraction efficiency (EE) of the fluid for the three target plants. The results of the GFE experiments are presented in Table 3.18.

Table 3.18. Arsenic extracted by GFE and 0.05 M-HCl. A pH of 1.8 was used for both extraction media. Arsenic speciation in GFE extracts was done by HPLC-HGAAS. Results are mean ± 1 s.d. from 3 independent determinations.

Sample ID	Total As ($\mu g.g^{-1}$)	As by GFE ($\mu g.g^{-1}$)	As by 0.05 M-HCl ($\mu g.g^{-1}$)	Species in GFE Extracts ($\mu g.g^{-1}$)				Sum of species in GFE
				As(III)	As(V)	MA	DMA	
P-1	241 ± 9	29.4 ± 2.4	55.5 ± 2.1	nd	28.4 ± 0.1	1.4 ± 0.1	nd	29.8 ± 0.2
P-3	146 ± 5	31.3 ± 5.1	77.6 ± 1.8	4.8 ± 0.1	26.8 ± 0.2	nd	nd	31.6 ± 0.3
P-6	24 ± 1	20.0 ± 0.8	14.3 ± 0.2	16.5 ± 0.4	6.1 ± 1.3	nd	nd	22.6 ± 1.7

The average (n = 3) extraction yields by GFE from P-1, P-3 and P-6 were 29.4 ± 2.4 $\mu g.g^{-1}$ (12% EE), 31.3 ± 5.1 $\mu g.g^{-1}$ (21% EE) and 20.0 ± 0.8 $\mu g.g^{-1}$ (83% EE), respectively, showing significant extraction of As from plants by gastric fluid. Although the extraction efficiency decreased with increasing total As in the plants, more than 80% of the total 24 $\mu g.g^{-1}$ was extracted from P-6 by gastric fluid. The extraction efficiencies of GFE were comparable to those of 100% water (see Fig 3.3).

The presence of MA in P-1 was confirmed by gastric fluid extraction (GFE) of the plants. The average (n = 3) amount of MA extracted from the plant P-1 was 1.4 ± 0.1 $\mu g.g^{-1}$ which compared favorably with 1.51 ± 0.04 $\mu g.g^{-1}$ extracted by the 1:1 water-methanol sonication method. Thus pepsin, and similar enzymes such as cellulase and CereCalase™ having matrix-modifying properties, shows potential for simultaneous analysis of the organic and inorganic arsenic in plant samples.

3.17. EXTRACTION BY A COMBINATION OF ACID AND ENZYME

In the gastric fluid extraction (GFE) described above (Section 3.15) the gastric fluid extracted a significant amount of arsenic as well as organoarsenic species MA from the plant. GFE showed improved extraction of arsenic over that of 1:1 water-methanol solvent (see Fig. 3.3). GFE did not require any solvent removal from the extracts unlike water-methanol mixture from which methanol had to be removed prior to analysis. Gastric fluid extracts were introduced directly to the ICP for total arsenic and into the HPLC column for arsenic speciation. Because of the advantages of gastric fluid over the water-methanol solvent, further investigation with gastric fluid for finding a potential solvent for the extraction of both inorganic and organic arsenic species was carried out. Extraction efficiency of solvents made of enzyme (pepsin), acid (HCl) and salt (NaCl) was explored.

3.17.1. The Enzyme in Various HCl Concentrations

Experiments were conducted to determine arsenic extraction efficiency of pepsin in HCl solutions with varying concentrations. Four 1.25 g.L^{-1} pepsin solutions were prepared in 0.05 M-, 0.1 M-, 0.25 M- and 0.5 M-HCl. A field horsetail (DL-207b) that showed the presence of MA from earlier analysis was extracted. Duplicate independent replica samples in 0.05 M- and 0.25 M-pepsin-HCl solutions and a blank for each solution were analyzed to ascertain reproducibility and quality assurance. The pH values of the HCl and pepsin-HCl solutions were measured. The pH values of the blank HCl solutions (i.e., without pepsin) were 1.50, 1.24, 0.94 and 0.70 and those with pepsin were 1.39, 1.18, 0.90 and 0.68, respectively. The extraction process was carried out in three steps with sonication as described in Section 3.3.2. The results of extraction experiments are reported in Table 3.19. Extraction results showed that efficiency of extraction did not change to a significant amount by changing HCl concentration from 0.05 M to 0.5M. Further experiments, therefore, were conducted using 0.1 M-HCl that was more commonly employed in this study and close in concentration to stomach acidity.

Table 3.19. Effect of HCl concentrations on extraction efficiency of gastric fluid.

HCl concentration	As extracted ($\mu g \cdot g^{-1}$)[a]
0.05 M	97.7 ± 0.3[b]
0.1 M	101
0.25 M	103 ± 2[b]
0.5 M	104

[a]The test plant was field horsetail (DL-207b) with total arsenic 122 $\mu g \cdot g^{-1}$.
[b]Mean ± ave. deviation from two independent determinations.

3.17.2. Effect of Pepsin Concentration and Salt on Extraction

The extraction of arsenic from plant sample was conducted using solvents with various pepsin concentrations, and pepsin and sodium chloride together in solution. Solutions containing 0.625, 1.25 and 2.5 g.L^{-1} pepsin in 0.1 M-HCl were prepared. The solution containing both pepsin and salt had 1.25 g.L^{-1} pepsin and 8.77 g.L^{-1} salt in 0.1 M-HCl. The DDW with 1.25 g.L^{-1} pepsin, blank 0.1M-HCl and blank DDW were also used as solvent in order to compare the extraction results as reported in Table 3.20. The increment of pepsin concentration in 0.1 M-HCl yielded insignificant improvement in the EE for As from the plant. However, a significant improvement in the EE from 101 ± 1 to 120 ± 4 was observed by the addition of salt to the 1.25 g.L^{-1} pepsin in 0.1 M-HCl solution. The EE of this solution was better than that obtained from 1.25 g.L^{-1} pepsin in DDW or 0.1 M-HCl/DDW alone. The latter solvents, however, extracted the same total amount of arsenic from the plant except that blank water extracted more As(III) and less As(V) than the water-pepsin solution. All of the solvents tested, however, extracted similar amounts (~8.0 $\mu g \cdot g^{-1}$) of MA from the plant. The test plant extracted by the solvents was the same field horsetail sample (DL-207b) as used in section 3.17.1.

Table 3.20. Effect of pepsin concentration with and without salt on extraction efficiency of gastric fluid. The test plant was field horsetail (DL-207b) with a total As content 122 µg.g^{-1}. Results are mean ± ave. deviation from 2 independent determinations.

Pepsin Conc. (g.L^{-1})	Solvent	As Extracted (µg.g^{-1})				Total As Extracted, µg.g^{-1}
		As(III)	DMA	MA	As(V)	
0.625[a]	0.1 M HCl	11.6	nd	8.8	88.6	109
1.25	0.1 M HCl	13.0 ± 0.3	nd	8.1 ± 0.5	80.2 ± 0.3	101 ± 1
2.5	0.1 M HCl	14.1 ± 0.3	nd	8.9 ± 0.4	87.2 ± 7.4	110 ± 8
1.25 +Salt[b]	0.1 M HCl	12.2 ± 2.2	nd	11.5 ± 1.0	96.2 ± 0.6	120 ± 4
Blank	0.1 M HCl	9.9 ± 0.2	nd	8.4 ± 0.1	84.2 ± 6.5	103 ± 7
1.25	DDW	16.5 ± 0.1	nd	7.6 ± 0.8	70.1 ± 0.5	94.2 ± 1.0
Blank	DDW	25.9 ± 0.6	nd	8.7 ± 0.8	59.7 ± 0.3	94.4 ± 1.7

[a] Duplicate sample was damaged-lost prior to speciation analysis.
[b] Salt concentration was 8.77 g.L^{-1}, nd = not detected.

This plant yielded high arsenic (i.e., high EE) in dilute HCl solvents (see Fig. 3.3) due possibly to its matrix composition. The matrix compositions of field horsetail are described in Chapter 4.10. However, variable EE's are expected from plants having various matrix components.

3.17.3. Extraction of Plants by Pepsin-HCl-Salt Solution (PHS)

The combination of pepsin and salt in 0.1 M-HCl was more efficient in extracting arsenic (see Table 3.20) and was employed to extract a number of Deloro plants for arsenic. The plants were extracted following the three-step extraction aided by sonication as described in Section 3.3.2. The results of pepsin-HCl-salt (PHS) extraction of the plants are reported in Table 3.21. The total EE from PHS determined by HPLC-HGAAS, ICPAES and that from the sequential method are given in the last three columns for comparison. Percent total extractions from the ICPAES and HPLC-HGAAS agree well. The PHS showed significant improvement over the sequential method for a number of plants. For plants DL-105 and DL-108, approximately 100% more arsenic was extracted and for DL-209 and DL-220, more than 200% more arsenic was extracted by the PHS method than the sequential method. The

EE's of DL-207b, the test plant (a field horsetail) used for the extraction experiments, were similar, as expected, by the PHS and sequential methods.

Table 3.21. Pepsin-HCl-salt (PHS) extraction of plant samples. EE's of PHS and sequential extractions are reported for comparison. Results are mean ± ave. deviation, n = 2 for plants DL-207b and DL-209.

Plant ID	Total As, $\mu g \cdot g^{-1}$	As(III), $\mu g \cdot g^{-1}$	DMA, $\mu g \cdot g^{-1}$	MA, $\mu g \cdot g^{-1}$	As(V), $\mu g \cdot g^{-1}$	Total from HPLC-HGAAS, $\mu g \cdot g^{-1}$	% EE of PHS by HPLC-HGAAS	% EE of PHS by ICPAES	% EE of sequential extraction
DL-105	8.4	4.6	nd	nd	4.3	8.9	106	111	49
DL-108	11.3	9.8	nd	nd	nd	9.8	87	100	50
DL-207	122	14.3 ± 1.0	nd	8.9 ± 0.3	96.5 ± 2.4	120 ± 4	98 ± 1	85 ± 1	85 ± 1
DL-209	20.3	15.4 ± 0.1	nd	nd	nd	15.4 ± 0.1	76 ± 1	69 ± 2	23
DL-220	18.7	13.8	nd	nd	nd	13.8	74	67	19

An inspection of the speciation results of PHS extractions above (Table 3.21) and sequential extractions in Chapter 4 (Table 4.11) revealed that the success of PHS lay in the capability of extracting more As(III) from the plants than the sequential method. In the sequential method, 1:1 v/v water-methanol solvent could not extract significant or detectable amount of As(III) from DL-105, DL-209 and DL-220, but PHS extracted substantial amount of arsenite. While single solvent extraction involving 0.1M-HCl yielded better results than the sequential method for some plants (Table 5.1), for DL-209 the overall EE of PHS was superior to that of the other methods. For example, in case of DL-209, EE attained by the sequential, single-solvent extraction by 0.1 M-HCl and PHS methods were 23, 57 and 69, respectively (Table 3.21 and Table 5.1). Since the efficiency of the PHS method is due to its capability to extract more arsenite, the least ionic or polar of the arsenic species, the presence of the organic enzyme pepsin in the PHS extractant might have contributed towards its higher extraction efficiency. It appeared, therefore, that PHS extraction might produce better EE for plants containing higher amounts of As(III). The disadvantage of PHS was evident from its inability to extract DMA

Arsenic in Plants: Extraction and Speciation

from the plants (DL-105, DL-108, DL-209 and DL-220; for details, see Chapter 4) but was detected in 1:1 water-methanol extracts. It also lacked the general EE required for a variety of terrestrial plants.

3.18. SEQUENTIAL ARSENIC EXTRACTION METHOD: DEVELOPMENT

3.18.1. Optimization of HCl-Sonication Extraction

Among the four acid solutions tested, 0.1 M-HCl and 0.05 M-HCl showed similar efficiencies (average EE > 90%) in extracting As from the target plant matrices (Fig. 3.3). Since 0.05 M-HCl was the milder of the two it was selected for optimization experiments. The effects of number of extraction steps, sample mass, time of sonication, size of extraction tube, and volume of the extractant on extraction efficiency (EE) were evaluated.

3.18.1.1. Schemes of Extraction: Flowchart

To determine the sequence of extraction of arsenic from plants by solvents a scheme of extractions was prepared. The schemes of extractions are depicted as a flowchart in Figure 3.15. Extraction experiments were carried out following the flowchart and the results compared. All extractions were carried out in three steps by solvent-sonication technique.

According to the Scheme A, Plant P-1 was extracted by 0.05 M-HCl and the residue was extracted by 1:1 water-methanol. In Scheme B, the plant was extracted first by 1:1 water-methanol and the residue was extracted by 0.05 M-HCl. In Scheme C, a solvent comprising 1:1 v/v 0.1 M-HCl and methanol was used and the plant was extracted without the residue being extracted further. For each extraction scheme, the plant was extracted in triple replicas and one sample blank was included.

The 1:1 water-methanol extracts were evaporated to dryness in an oven and reconstituted in DDW. All extracts were analyzed by ICPAES for total arsenic and arsenic speciation was conducted by HPLC-HGAAS. The results of sequential scheme extractions are reported in Table 3.22. The amounts of arsenic extracted by 0.05 M-HCl in both A and B Schemes were much higher than that extracted by 1:1 water-methanol solvent indicating better extraction capability of dilute HCl than the water-methanol combination. The total amounts of arsenic extracted by schemes A and B were 190 ± 20 µg.g^{-1} and 152 ± 17 µg.g^{-1}, respectively. The Student's t-test for paired samples showed no significant difference (at 95% CL) between the mean EE's of the Schemes A and B. Thus the sequence of whether HCl acid or water-methanol was used first did not make a significant difference on the total arsenic extracted from the plant.

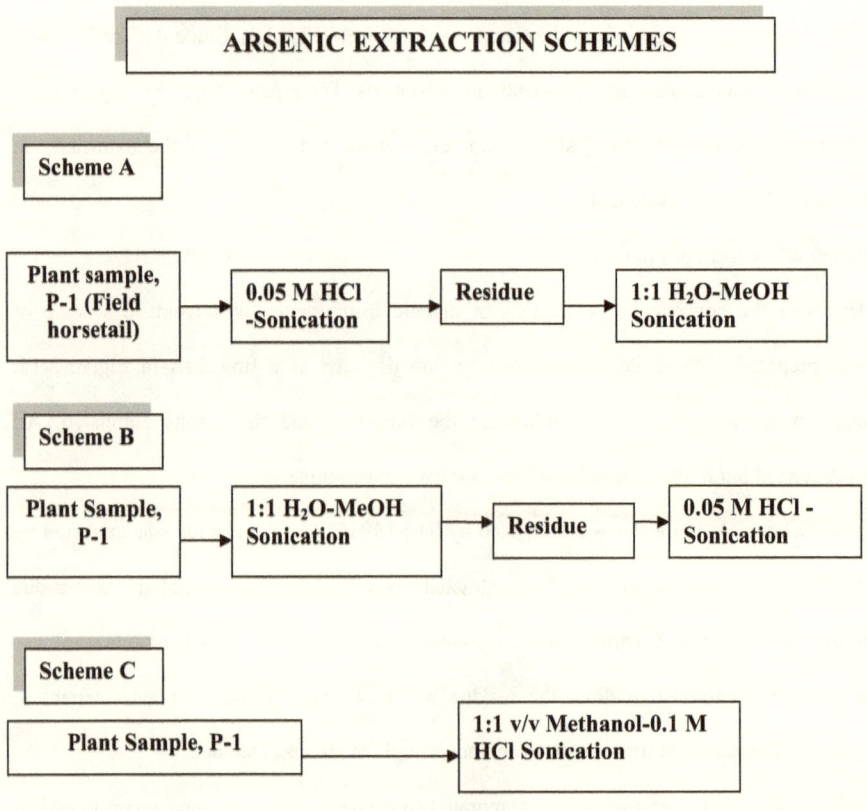

Figure 3.15. Extraction schemes used for the optimization of the arsenic extraction methods are shown.

In scheme C, the quest to see if both organic and inorganic species could be determined in one solvent, extractions of P-1 (241 µg.g^{-1} total As, 1.5 µg.g^{-1} MA) were performed in 1:1 v/v 0.1M-HCl/methanol solvent aided by sonication. The average (n =3) yield of total arsenic determined by ICPAES was 69.2 ± 1.6 µg.g^{-1}. This yield was much lower than 202 ± 6 µg.g^{-1} extracted by 0.05 M-HCl-sonication, but higher than 7.7 ± 1.0 µg.g^{-1} extracted by 1:1 water-methanol (Fig. 3.3). Although the strength of acid in the latter solvent was 0.05 M,

Table 3.22. Sequential schemes and 1:1 methanol-0.1 M-HCl extraction results. A and A' are sequential extractions of Scheme A; B and B' are of Scheme B; and C and C' are of Scheme C from Figure 3.15. Plant sample was P-1 (field horsetail).

Scheme	Replicates	As by ICP-AES, µg.g^{-1}	Total from each scheme, µg.g^{-1}	Average ± 1 s.d., n = 3; µg.g^{-1}
Scheme A	A-1	185	199	190 ± 20
	A-2	188	203	
	A-3	152	167	
	A-4-Blank	nd		
	A'-1	14.5		
	A'-2	14.8		
	A'-3	14.8		
	A'-4-Blank	nd		
Scheme B	B-1	7.36	141	152 ± 17
	B-2	7.62	144	
	B-3	7.27	172	
	B-4-Blank	nd		
	B'-1	134		
	B'-2	137		
	B'-3	165		
	B'-4-Blank	nd		
Scheme C	C-1	70.9	70.9	69.2 ± 1.6
	C-2	69.0	69.0	
	C-3	67.7	67.7	
	C-4-Blank	nd		

its efficiency was drastically reduced by the presence of methanol. The plant (P-1, field horsetail) was confirmed earlier to have MA in the matrix by 1:1 water-methanol (see Table 3.9 Table 3.18). In the sequential extraction schemes, it was again detected only in 1:1 water-methanol extracts of scheme B by speciation analysis with HPLC-HGAAS. Speciation analysis by HPLC-HGAAS of 1:1 v/v mixture of 0.1 M-HCl-methanol sonication extracts of P-1 did not detect MA in the extract. This shows either suppression of extraction and/or analytical difficulty of organic species in the HCl medium in HPLC-HGAAS determinations. Similarly, the very low yield of total arsenic with methanol that was found earlier (Fig. 3.2 and Fig. 3.3) showed methanol's suppression of extraction of the inorganic arsenic from plants.

3.18.1.2. Effect of Number of Extraction Steps, Sample Mass, Volume/Sample Mass Ratio, Size and Shape of Extraction Vessel and Sonication Time on EE

In the beginning, five steps were carried out for extracting arsenic from plant matrices. In the extraction experiments reported to this point in the thesis, 0.5 g of the dried and ground plant materials were used as sample mass. Sonication was carried out in 5 steps in 10 mL of extraction solvent placed in 15 mL size extraction tubes. Since 0.05 M-HCl was the milder of the two it was chosen for optimization experiments. Experiments were carried out to determine the extraction output for each step. Extractions with sample masses 0.25 g and 0.10 g were carried out to determine the effect of mass on EE. Volume to mass ratios (mL/g) of 40, 150 and 200 were evaluated for EE. The effects of sample mass, number of extraction steps, time of sonication, size of extraction tube, and volume of the extract on extraction efficiency (EE) were evaluated. All samples were weighed accurately up to $\pm$ 0.1 mg level.

The results of optimization experiments are presented in Table 3.23, Table 3.24 and Table 3.25. Total EE from all steps in Table 3.23 amounted to $93 \pm 3\%$ while that from the first two steps amounted to $79 \pm 2\%$ indicating extraction of the most of arsenic from P-6 within the first few steps. The effect of sample weight on EE was observed (Table 3.23 and Table 3.24). A switch of sample mass from 0.50 g to 0.25 g of P-6 increased EE from $59 \pm 1\%$ (in Table 3.23) to $93 \pm 3\%$ (in Table 3.24) within the first step. Similar effects were observed for the other two plants after sample masses had been reduced from 0.25 g to 0.1 g. EE in the first step increased from $36 \pm 1\%$ to $68 \pm 3\%$ and $48 \pm 4\%$ to $66 \pm 2\%$ for P-1 and P-3, respectively. Total

EE's of two steps with lower sample weight (0.1 g) were found to be better than that from two steps with higher sample weight (0.25 g) for P-1 and P-3, the higher arsenic containing plants.

Table 3.23. Step by step extraction of As from plant P-6 (Field horsetail, 24 $\mu g.g^{-1}$ total As) by 0.05 M-HCl-sonication process. Sample was 0.5 g (± 0.1 mg). Results are mean ± 1 s.d. from 3 independent determinations.

Extraction steps	Step 1	Step 2	Step 3	Step 4	Step 5	Total
Time of sonication	20 min	10 min	10 min	10 min	10 min	60 min
Amount of arsenic extracted ($\mu g.g^{-1}$)	14.2 ± 0.3	4.9 ± 0.2	1.9 ± 0.2	0.92 ± 0.03	0.48 ± 0.01	22.4 ± 0.7
EE (%)	59 ± 1	20 ± 1	7.9 ± 0.8	3.8 ± 0.1	2.0 ± 0.1	93 ± 3

Table 3.24. Effect of sample weight on extraction efficiency (EE). Time of sonication for each step is indicated. Results are mean ± 1 s.d. from 3 independent determinations.

Plant ID	Total As in plant ($\mu g.g^{-1}$)	Sample weights, (g)	Extraction efficiency (%)			Total EE (%)
			Step 1 (20 min)	Step 2 (10 min)	Step 3 (10 min)	
P-1	241	0.10	68 ± 3	9.1 ± 0.6	--	77 ± 4
		0.25	36 ± 1	28 ± 2	8.8 ± 0.1	73 ± 3
P-3	146	0.10	66 ± 2	13 ± 1	--	79 ± 3
		0.25	48 ± 4	24 ± 4	7.8 ± 0.1	80 ± 8
P-6	24	0.10	86 ± 3	11 ± 3	--	97 ± 6
		0.25	93 ± 3	17 ± 4	4.2 ± 0.1	114 ± 7

Optimization experiments were conducted in one step with 30 minutes of sonication and varying the ratio of solvent volume to sample weight. Incrementing this ratio from 40 to 200 did not increase EE to a significant amount (Table 3.25).

Table 3.25. Effects of ratio of extractant volume to sample mass, extractant volume, and size of extraction tube on extraction efficiency (EE). One-step 30 minutes and one-step 20 minutes (as indicated) of sonication were employed. Results are mean ± 1 s.d. from three independent determinations.

Plant ID	Size of extraction tube	Volume of extraction solution (ml)	Mass of sample (g)	Volume to mass ratio	EE (%)
	15 ml (1.5 cm dia.)	10	0.25	40	56 ± 2
P-1	50 ml (3 cm dia.)	30	0.20	150	57 ± 1
	50 ml (3 cm dia.)	40	0.20	200	59 ± 4
P-3	15 ml (1.5 cm dia.)	10	0.25	40	76 ± 3
P-6	15 ml (1.5 cm dia.)	10	0.25	40	87 ± 3
	15 ml (1.5 cm dia.)	10	0.25	40	84 ± 4
P-6[a]	50 ml (3 cm dia.)	10	0.25	40	93 ± 3
	50 ml (3 cm dia.)	20	0.25	80	91 ± 3

[a] One step 20 minutes sonication extraction.

Further experiments were conducted to determine effect of tube size and solvent volume on extraction with 0.25 g of P-6 placed in 15 mL and 50 mL extraction tubes and 0.05 M-HCl solution as extractant: 10 mL in 15 mL extraction tube, 10 mL and 20 mL in 50 mL extraction tubes. These results are reported also in Table 3.25. Extraction was done in one step with 20 minutes of sonication for all samples. An EE of 84 ± 4% was found in 15 mL tubes and 93 ± 3% in 50 mL tubes with 10 mL of solution in each tube showing an 8-10% improvement in EE from 15 ml tube to 50 ml tube. However, a 91 ± 3% EE was observed for 20 mL solutions indicating no significant improvement due to change of extractant volume.

3.18.2. The Development and Optimization of Sequential Method: Conclusions

The experimental results showed that HCl had significantly affected extraction and analysis of organoarsenic species from plants in 1:1 water-methanol medium. On the other hand, methanol severely decreased EE of HCl when mixed together. The sequence of acid extraction prior to water-alcohol also caused difficulty in the speciation of the organoarsenic species by HPLC-HGAAS. Conversely, the water-alcohol extraction prior to the acid extraction did not cause significant difference in the extraction of the total extractable arsenic by the sequential method and the extraction by the solvents independently. The lack of difference may be mainly due to the very small quantity of the organoarsenic species compared to inorganic arsenic species found in the plants grown on highly contaminated soils. Nonetheless, the sequential method of 1:1 water-methanol extraction followed by 0.05 M-HCl provided scope for analyzing organic species without interference of the predominant inorganic arsenic in the plants. Moreover, the sequential method significantly improved sample handling including sample consumption and analyst time.

Further optimization of the method showed that three steps rather than the usual five steps of solvent sonication extraction would suffice for the quantitative extraction of arsenic from most plants. This optimization also saved time and reagent. Decreasing the sample mass by half from 0.5 g to 0.25 g improved EE to a significant extent in the solvent sonication methods. The improvement is understood in light of the fact that the less amount of sample mass reduced the physical barrier to the sonication process paving the way for better EE of the analyte. Less amount of sample might also have improved the solubility of analyte in the extractant solvent. The choice of sample mass, however, would also depend on the analyte concentration of the original sample, its extractability, and the sensitivity of the detection instrument employed for analysis. The results of our study indicated that the size and shape of the extraction vessels might have influenced the EE of arsenic in the solvent-sonication processes.

The arsenic speciation experiments have shown the presence of both inorganic and organic arsenic in plant tissues. Though the inorganic As(III) and As(V) are the predominant species in terrestrial plants, MA and DMA are also detected in lower concentrations.[96,132,133] Other organic arsenic compounds such as arsenobetaine, arsenocholine, and different arsenosugars have also been detected in plant tissues.[92,105,139] Therefore, it is important to extract both inorganic and organic arsenic to obtain a broader picture of the species in plants. At the same time, in order to keep the integrity of the species, milder extractants must be used.[26,27,98,99] Although the water-methanol mixture extracted both organic and inorganic arsenic compounds from plants and biological tissues, its capability for extracting inorganic arsenic was inferior to that of HCl or even water (see Fig. 3.2 and Fig. 3.3). On the other hand, HCl was efficient in extracting inorganic arsenic but detection of the trace organoarsenic species was difficult in this medium as discussed in Section 3.15.

The difficulty of analyzing the organoarsenic species in HCl medium might have been due to the presence of large concentrations of inorganic arsenic and other polar and ionic plant materials co-extracted in the acid medium. Nonetheless, the use of either water-methanol or HCl alone would result in an incomplete extraction or detection of the arsenic species from plants. Therefore, the separation of the trace organic species from the inorganic arsenic in alcohol media would be desirable for their determination. This is also desirable because of the fact that As in chloride media suffers both spectral and non-spectral interferences in mass spectrometric analysis.[140]

During the optimization of the sequential method, extraction of P-1 (*Equisetum arvense*) with 1:1 water-methanol followed by 0.05 M-HCl, and the single solvent extraction with 0.05 M-HCl alone, yielded similar amounts of arsenic by both methods. The difference was insignificant due to the very low concentrations of the organoarsenic compounds compared to those of the inorganic arsenic. Overnight (~20 hours) soaking of a number of terrestrial plants in 0.1 M-HCl and 1:1 water-methanol did not improve EE of the solvents for the plants. On the contrary, the extraction by acid prior to the extraction by 1:1 water-methanol was found to be detrimental to the extraction/determination of organoarsenic species in plants. Therefore, the extraction sequence of water-methanol followed by the acid was adopted. Moreover, the sequential extractions and batch drying of the water-alcohol extracts in an oven significantly reduced sample handling and consumption, and analyst time.

The primary goals of the current investigation were to develop arsenic extraction and speciation methods that would be suitable for analyzing many plant samples such as those in risk assessments using common laboratory methods, reagents and analytical instruments. The instrumental techniques used in the analysis of arsenic in this study were

HGAAS, ICPAES, HPLC-HGAAS and HPLC-ICPMS. Hydride generation (HG) method has been proved sufficiently sensitive for the analysis of arsenic in environmental samples.[3,27]

Although ICPMS has become the method of choice for many analytical laboratories, its advantage over HGAAS for the determination of arsenic was not established by comparative studies.[74,75,141] Besides, procurement, consumables and maintenance costs of ICP-MS instrument are much higher than those for the HG-AAS instrument and would be disadvantageous for commercial laboratories analyzing many environmental samples.[76]

3.19. SEQUENTIAL ARSENIC EXTRACTION METHOD: APPLICATION TO REAL SAMPLES

A number of plants from the arsenic contaminated areas in Yellowknife in NWT, Canada were extracted by the sequential method. Also, many plants comprising a variety of species collected from Deloro during the year 2002 and subsequent years were extracted by the sequential method for arsenic speciation. The results of extraction and speciation of the Deloro plants will be discussed in detail in Chapter 4 and Chapter 5. The results of extraction and speciation of arsenic from the Yellowknife plants are presented here. The detailed Sequential Extraction Method is given in Appendix A.

The sequential extractions were carried out with 1:1 water-methanol followed by 0.1 M-HCl for fifteen Yellowknife plants and the results are reported in Table 3.26. Though the optimization experiments were carried out with 0.05 M-HCl, which was as efficient as 0.1 M-HCl, subsequent extractions were done in 0.1 M-HCl since the overall reproducibility of analysis was found to be better in 0.1 M-HCl than that in 0.05 M-HCl.

The total arsenic concentrations in these plants were determined by ICPAES following dry ashing and acid digestion. Total arsenic concentrations ranged between 1-95 $\mu g.g^{-1}$ and are reported in the same Table (Table 3.26). Most of these plants were grass species; the grasses are usually silica-rich, have hard-tissue and are difficult to process.[142] This difficulty was also reflected in the 1:1 water-methanol extraction of the plants whose EE ranged between 4-89% with a median at about 29%.

Kalam Mir

The sequential extraction followed by 0.1 M-HCl extracted more arsenic from most of the plants listed in Table 3.26. Had the extractions been carried out only with 1:1 water-methanol, the commonly used solvent, low extraction efficiency would have been achieved. From the plants that were mostly grasses, the overall EE from sequential extractions ranged between 27-128% with a median at 46% that was almost double of the median EE of 1:1 water-methanol extractions.

Table 3.26. Extraction of arsenic from Yellowknife plants by sequential extraction method. Extraction efficiencies of 1:1 water-methanol (traditional method of plant extraction) and sequential method are compared. Results are mean ± ave. deviation from two independent analyses for a number of plants are reported for quality assurance.

Plant ID	Latin name	Common name	Total As in plant	As in 1:1 Water-MeOH, ($\mu g \cdot g^{-1}$)	As in 0.1 M-HCl, ($\mu g \cdot g^{-1}$)	Sum of two extracts, ($\mu g \cdot g^{-1}$)	% EE of 1:1 water-MeOH	% EE of sequential method
YK-1	Agropyron trachycaulum	Slender wheat grass	95.2	8.2	20.2	28.4	9	30
YK-2	Epilobium angustifolium	Fireweed	4.6	1.5	nd	1.5	33	33
YK-3	Archtostaphytos uva-ursi	Common bearberry	18.5	1.8	4.8	6.6	10	36
YK-4	Agropyron trachycaulum	Slender wheat grass	8.0	1.5	1.5	3.0	19	38
YK-5	Senecio vulgaris	Common groundsel shoots	38.5	5.8	20.6	26.4	15	69
YK-6	Agrostis scabra	Rough hair grass shoots	47.5	24.9 ± 0.8	25.0 ± 0.8	49.9 ± 1.6	52 ± 2	105 ± 3
YK-7	Picea mariana	Black spruce bough/cones	12.6	0.78 ± 0.20	2.8 ± 0.1	3.6 ± 0.3	6.2 ± 1.5	29 ± 2

(Table continued)

Table 3.26. (Continued)

Plant ID	Latin name	Common name	Total As in plant	As in 1:1 Water-MeOH, ($\mu g.g^{-1}$)	As in 0.1 M-HCl, ($\mu g.g^{-1}$)	Sum of two extracts, ($\mu g.g^{-1}$)	% EE of 1:1 water-MeOH	% EE of sequential method
YK-8	*Hordeum jubatum*	Foxtail barley	4.4	1.2	nd[1]	1.2	27	27
YK-9	*Calamagrostis canadensis*	Blue joint-1	4.0	2.5	nd	2.5	63	63
YK-10	*Calamagrostis canadensis*	Blue joint-2	8.1	3.7 ± 0.2	nd	3.7 ± 0.2	46 ± 2	46 ± 2
YK-11	*Calamagrostis canadensis*	Blue joint-3	77.7	3.4 ± 0.4	20.2 ± 1.1	23.6 ± 1.5	4.4 ± 0.5	30 ± 2
YK-12	*Calamagrostis canadensis*	Blue joint-4	3.8	1.1	1.3	2.4	29	64
YK-13	*Calamagrostis canadensis*	Blue joint-5	3.0	1.4	1.7	3.2	48	105
YK-14	*Hordeum jubatum*	Foxtail barley	0.90	0.80	nd	0.80	89	89
YK-15	*Epilobium angustifolium*	Fireweed	2.1	1.1	1.5	2.6	52	128

[1] nd = not detected.

Table 3.27. Speciation of arsenic in the plant extracts by HPLC-HGAAS (see Table 3.26). Total As in the extracts was determined by ICP-AES. Mean ± ave. deviation from 2 independent determinations are reported for plants YK-6 and YK-11. Concentrations are in $\mu g.g^{-1}$ and nd = not detected.

Plant ID	As(III)		DMA		MA		As(V)		Sum of species	Total by ICPAES
	1:1 Water-MeOH	0.1 M-HCl	1:1 Water-MeOH	0.1 M-HCl	1:1 Water-MeOH	0.1 M-HCl	1:1 Water-MeOH	0.1 M-HCl		
YK-1	3.1	nd	nd	nd	0.36	nd	4.3	19.5	27.3	28
YK-3	nd	nd	nd	nd	nd	nd	1.0	5.6	6.6	6.6
YK-4	0.53	nd	nd	nd	0.23	nd	0.83	2.1	3.6	3.0
YK-5	1.5	7.6	nd	nd	nd	nd	3.4	20.9	33.4	26
YK-6	9.4 ± 1.4	0.49 ± 0.13	nd	nd	nd	nd	11 ± 1	29 ± 2	50 ± 5	50
YK-11	0.50 ± 0.11	0.49 ± 0.08	nd	nd	0.25 ± 0.02	nd	2.1 ± 0.2	24 ± 1	27 ± 1	24

The 1:1 water-methanol extracts were dried and regenerated in a smaller volume of DDW for arsenic speciation. The extracts from the sequential extractions of a number of plants were analyzed by HPLC-HGAAS and the results are reported in Table 3.27. The results of speciation show the presence of mostly inorganic species, As(III) and As(V), in the plants. The predominance of As(V) over As(III) in the extracts of the Yellowknife plants was observed to be similar to that in the extracts of the test plants from Deloro discussed earlier (Fig 3.10). The organoarsenic species MA was detected in 1:1 water-methanol extracts of plants YK-1, YK-4 and YK-11.

In order to know the fate of the unextracted portion of total arsenic of the Yellowknife plants, mass balance experiments by acid digestion of the residues of sequential extraction was carried out. The mass balance accounted for all unextracted arsenic in the plant residues (Table 3.5). The arsenic in the residue is considered to be the arsenic not extractable by the sequential method from these plants. The knowledge of chemical and physical states of the unextracted arsenic may provide insight into understanding the factors that influence extraction of arsenic from the plant matrix. It is generally believed that the low EE was due to a number of factors such as insoluble forms of arsenic, the chemical and/or physical bonding of As to the plant matrix, and trapping of arsenic compounds inside the plant vascular tissues.[92,102] The XANES experiments, as mentioned earlier, provided results that corroborated our findings of the arsenic species in solid plant matrices and the liquid plant extracts. The results of further extractions and XANES analysis will be discussed in detail in Chapters 4 and 5.

CHAPTER 4: ARSENIC EXTRACTION AND SPECIATION IN PLANTS FROM DELORO, ONTARIO, CANADA

4.1. INTRODUCTION

This research project involves the extraction and speciation of arsenic in plants grown on highly impacted areas. The initial work used plant samples collected from Deloro in November 2001. In addition, many plant samples were collected from Deloro throughout the year 2003 and in 2004, and analyzed for total arsenic and the species as well.

The development of the sequential method of extraction of arsenic from plant matrices has been detailed in Chapter 3. The results of initial determination of arsenic in the plant and soil samples of Deloro have been given. The optimization of the sequential method, GFE and enzyme- acid -salt (PHS) extractions of plants have been described. The practical application of the sequential method to real samples has been demonstrated by extracting a number of terrestrial plants from Yellowknife, ON, Canada.

In this chapter, arsenic levels and species in plants and soil samples collected from Deloro are presented. Fresh knowledge on the dependence of extraction efficiency on total arsenic and plant species is reported. The results of uptake and seasonal variations in the concentrations of arsenic species in plants are summed up.

4.2. THE DELORO SITE

Deloro looks as scenic as other southern Ontario small towns except when a closer look is given at the abandoned gold mining areas. An aerial view of the area is shown in the photograph, Photo 4.1. Though the Ontario government has been working on covering and containing the open arsenic waste, tailing piles still remain. Occasional carcasses of migrating geese could be found near the arsenic tailing ponds (see Photos 3.4 and 4.3) despite the effort to keep them away using overhead wire lines (in Photo 3.4). The leaching of arsenic by rain and underground fountain waters is visible (Photo 4.2 and Photo 4.3). Dead vegetation is evident in some areas (Photo 4.4). After taking control of the site in 1979, the Ontario Ministry of Environment (MOE) took specific and significant actions to clean up the site.[57]

Arsenic in Plants: Extraction and Speciation

Photo 4.1. An aerial view of the historical Deloro gold mining area is shown. (Courtesy Ministry of Environment (MOE), ON, Canada).

Photo 4.2. A view of leached water from the 'covered' arsenic tailings flowing down the stream to form shallow pond shown in Photo 4.3.

Photo 4.3. A view of a shallow water body formed by leached water and rain water run-off over the arsenic tailings around the gold mining sites.

Photo 4.4. A view of dead vegetation strewn by the stream running with arsenic contaminated water.

Plants and wildlife are clearly exposed to arsenic in these areas. In spite of the adverse scenario, the opportunities for research on arsenic and other coexisting toxic elements and their impact on the environment have been created. The Deloro area may provide an ideal site for this to the scientific community.

4.3. SAMPLE COLLECTION AND PREPARATION

At the beginning of the project, soil and plant samples were collected from the historic and abandoned gold mining sites at Deloro, Ontario on November 15, 2001 through an arranged guided tour of the site by the Ontario Ministry of Environment. A team of two, comprised of Dr. John S. Poland and Kalam A. Mir, visited the Deloro site for the first time.

Over the years 2001-2004, many plant samples were collected from the arsenic contaminated areas of Deloro. Plant sample collection and preparation methods are described in Chapter 2. A map of the plant sample collection sites at Deloro is presented in Map 4.1. Sites 1, 2 and 3 were between the northwestern edges of the Young's Creek and southeastern corner of the tailings area. Sites 4, 5 and 6 were at the north of the industrial area and around the main historical mine area east of the Deloro village. Site 7 was at the south of the remote mine area on the west bank of the Moira river near Highway 7 and visited only once during the first exploration of Deloro for plant and soil samples.

Because the first sampling was at the end of the growing season in November, plants were either dried or decayed. Four soil and six plant samples were collected initially from different sites at the gold mine area. One soil (No. 1) and three plant samples (P-1: field horsetail, P-2: cattail and P-3: alkali grass; see Table 3.1) were collected from a meadow marsh area of high arsenic content (Site 1). The high arsenic concentration is due to the runoff from the arsenic trioxide tailing piles collecting in shallow ponds over the years. A fern sample (P-4) corresponding to soil sample No. 2 was collected from the high edge of a marsh of Young's Creek (north of Site 1). A sample of smooth horsetail (P-5) and soil sample (No. 3) was collected from a plot edging the rock barren bank of the Moira River (Site 7) south of the gold mine site near Highway 7. A second sample of the smooth horsetail (P-6) and soil sample No. 4 was collected from a capped tailing of calcium arsenite material (near Sites 4 and 6). These plant and soil samples are listed (Table 3.1) and analysis described in Chapter 3.

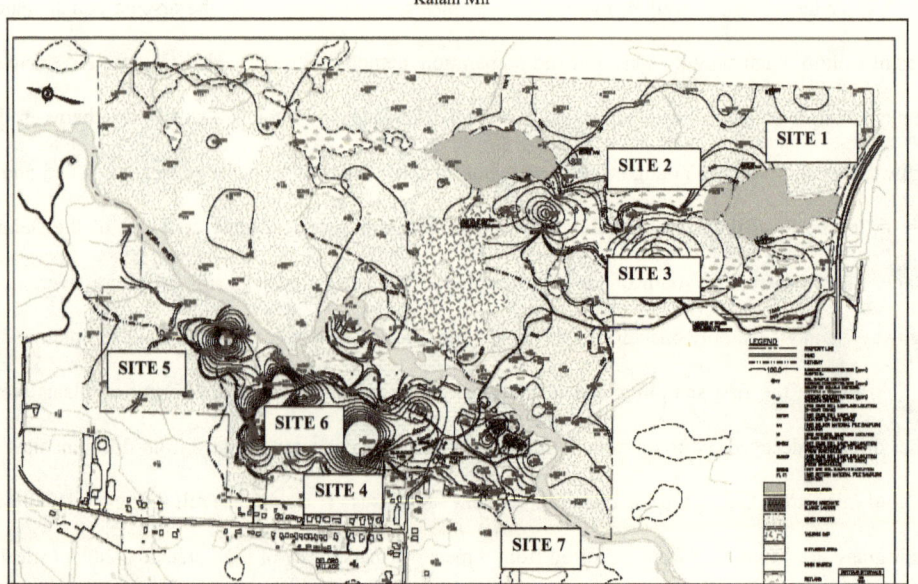

Map 4.1. Map shows Deloro gold mining area (Courtesy Ministry of Environment (MOE), ON, Canada). Plant and soil sampling sites are shown in boxes. The legends became unclear due to compression to fit a page of the thesis.

Arsenic in Plants: Extraction and Speciation

The initial results of total arsenic in soils and plants, and the amounts of arsenic extracted from the plants by different methods were significant. More plant species were needed for further experiments. In phase two of the project, soil and plant samples were collected from arsenic contaminated areas of Deloro in May, July and September of 2003. A total of 128 plant, moss and algae, comprising 74 different plant species, and six soil samples were collected from six sites as shown on the site map (Map 4.1). During this visit, a botanist was part of the team so that the identification of plants on site was possible. Two experts on the plant identification (Mr. Jakob Mueller and Mrs. Fiona Poland) were included in the team of five along with Kalam A. Mir, Dr. Allison Rutter and Dr. John S. Poland. All plants are listed with scientific and common English names in the Table 4.1. Representative pictures of collection and preparation of the plant samples from Deloro are shown in Photos 4.5-4.8.

Further collection of plant and soil samples from Deloro was carried out on August 31, 2004. In order to determine sampling variations within plant and soil samples, and any correlation between soil and plant arsenic concentrations, five plant and five soil samples were collected from three plots at three different locations. Approximately one square meter plots were marked at each of the sites Site-2, Site-5 and Site-6. Soil samples were collected as close to the corresponding plant samples as possible in each plot. Field horsetail, Canada goldenrod, and smooth horsetail samples were collected from Site-2, Site-5, and Site-6, respectively. Five plant samples collected from each site were washed, dried and ground individually.

4.4. TOTAL ARSENIC IN PLANTS AND SOILS

The methods for the determination of total arsenic in plants and subsequent development of the methods for its extraction and speciation have been described in Chapter 2 and Chapter 3, respectively. The total arsenic in plants was determined as the first step of all other determinations. The results of total arsenic for the three batches of plants are reported in Table 4.1. The Latin names (*italics*) as well as the common English names of the plants are given. The English names are mostly used throughout the thesis for discussion. Many plants that were collected in July were absent or too small to be collected in May due to early time of the growing season. Because of the optimum growth time of the season, more plant species were collected in July and, where possible, species from the May collection that showed total arsenic content 2 $\mu g.g^{-1}$ or more were resampled from the same site. Close to the end of the growing season in September, the plant species, which had been previously sampled and showed high levels of arsenic, were targeted. The total arsenic content in plants of the three batches ranged between <1 $\mu g.g^{-1}$ and >500 $\mu g.g^{-1}$ (Table 4.1). The relative average deviations (= 100x average deviation/mean) of duplicate analyses of total arsenic in plants of the three batches, ranged between 0.4 and 5.0 indicating good accuracy of the analytical results.

Arsenic in Plants: Extraction and Speciation

Table 4.1. Total Arsenic in Deloro plants determined by ICPAES after acid digestion is given. A number of plants were analyzed in duplicate and results are listed as (mean ± ave. deviation). Plants collected in May were given one/two digit ID Nos.; in July and September were given three digit, 100 and 200 ID Nos., respectively.

Plant ID (DL-#)	Latin name	Common name	Collected May 22, 03	Collected July 17, 03	Collected Sept. 18, 03
			Total As ($\mu g \cdot g^{-1}$)		
First site					
1	Cornus canadensis	Bunchberry	1.52		
2/102	Vaccinium angustifolium	Low blueberry	2.94	3.6	
3	Maianthemum canadensis	Wild lily of the valley	2.4		
4	Amelanchier stolonifera	Service berry	0.92		
5/106/200	Fragaria virginiana	Wild strawberry	2.35 ± 0.01	10.7	10.4
6/104/201	Onoclea sensibilis	Sensitive fern	8.6	22.1 ± 1.0	48.4
7	Typha sp.	Cattail	1.4		
8	--	White flower	0.6		
9	Alnus incana ssp. rugosa	Speckled alder	0.8		
10	Cornus stolonifera	Red osier dogwood	0.6		
11	Rumex acetocella	Sheep sorrel	0.6		
12	Lonicera tartarica	Tartarian honeysuckle	6.4		
13	Taraxacum officinale	Dandelion	1.2		
14/101	Equisetum arvense	Field horsetail (on land)	295	410	
15	Equisetum arvense	Field horsetail (in water)	532 ± 9		
16	--	Low shrub	0.8		
103/202	Pteridium aquilinum	Bracken fern		11.2	5.9
105/203	Verbena hastata	Blue vervain		8.4	8.5

(Table Continued)

Table 4.1 (continued)

Plant ID (DL-#)	Latin name	Common name	Collected May 22, 03	Collected July 17, 03	Collected Sept. 18, 03
			\multicolumn{3}{c}{Total As ($\mu g \cdot g^{-1}$)}		

Plant ID (DL-#)	Latin name	Common name	Collected May 22, 03	Collected July 17, 03	Collected Sept. 18, 03
First site (contd.)					
108	*Spirea alba*	Meadowsweet		11.3	
109	*Lythrum salicaria*	Purple loosestrife		11.2	
110	*Scirpus cyperinus*	Wool-grass		13.2	
111	*Carex sp.*	Sedge		3.5	
112	*Apocynum androsaemifolium*	Spreading dogbane		4.2	
113	*Rubus allegheniensis*	Common blackberry		1.6	
114	*Bromus inermis*	Smooth brome grass		1.9	
115	*Ambrosia artemisiifolia*	Common ragweed		9.1	
Second site					
17	*Populus sp*	Poplar	0.5		
18/119/206	*Solidago canadensis*	Canada goldenrod	3.8	5.8	1.1
19	*Chrysanthemum leucanthemum*	Ox-eye daisy	4		
20/121/205	*Aster lanceolatus*	Panicle aster	1.9	9.6	2.9
21/118/204	*Equisetum arvense*	Field horsetail	15	19.7 ± 0.2	19.4
22	*Trifolium hybridum*	Alsike clover	1.6		
116	*Puccinellia sp.*	Alkali grass		3.9	
117	*Puccinellia sp.*	Alkali grass		3.8	
120	*Typha sp.*	Cattail		1.5	
122	*Asclepias syriaca*	Common milkweed		3.3	
123	*Onopordum acanthium*	Scotch thistle		1.9	

(Table Continued)

Table 4.1 (continued)

Plant ID (DL-#)	Latin name	Common name	Collected May 22, 03	Collected July 17, 03	Collected Sept. 18, 03
			Total As ($\mu g.g^{-1}$)		
Third site					
26	*Tragopogon pratensis*	Goat's beard	1.4		
27	*Equisetum laevigatum*	Smooth horsetail	35.7		
29	*Bettula papyrifera*	Paper birch	0.4		
30	*Populus sp.*	Poplar	1.1		
31	*Salix*	Willow	1.1		
32	*Pinus strobus*	White pine	6.3		
33	*Acer*	Maple	0.9		
34/208	*Solidago canadensis*	Canada goldenrod	0.9		9.3 ± 0.1
60/124/207a, 207b[1]	*Equisetum arvense*	Field horsetail	24.9 ± 0.6	59.5 ± 3.0	157 ± 4, 122
125	*Puccinellia sp.*	Alkali grass		5.3	
126	*Hypericum perforatum*	St. John's wort		3.4	
Fourth site					
35	*Picea sp.*	Spruce	0.8		
36	*Lonicera tartarica tartarian*	Honey suckle	1.5		
Fifth site					
38	*Family Juncaceae*	Rush	1.4		
39	*Euthamia graminifolia*	Grass-leaved goldenrod	2.5		
40	*Populus balsamifera*	Balsam poplar	1.5		
41/128/209	*Lythrum salicaria*	Purple loosestrife	2.3	18.6	20.3
42/132	*Typha sp.*	Cattail	0.5	0.9	
43	*Amelanchior stobnifera*	Serviceberry	0.9		

(Table Continued)

Table 4.1 (continued)

Plant ID (DL-#)	Latin name	Common name	Collected May 22, 03	Collected July 17, 03	Collected Sept. 18, 03
			Total As ($\mu g \cdot g^{-1}$)		
Fifth site (contd.)					
44	*Querous maerocarpa*	Bur oak	0.6		
45/210	*Onoclea sensibilis*	Sensitive fern	3.27 ± 0.04		45.5
46	*Erythronium sp.*	Trout lily	3.1		
127	*Scirpus cyperinus*	Wool-grass		0.4	
129/211	*Solidago canadensis*	Canada goldenrod		8.2 ± 0.4	12.7
130/215	*Equisetum arvense*	Field horsetail		210	178
217	*Equisetum arvense*	Field horsetail			129 ± 4
131	*Puccinellia sp.*	Alkali grass		5.5	
133	*Chrysanthemum leucanthemum*	Ox-eye daisy		7.2	
134	*Aster lanceolatus*	Panicle aster		8.9	
135/212	*Equisetum laevigatum*	Smooth horsetail		36.8 ± 0.6	42
213	*Equisetum laevigatum*	Smooth horsetail			10.5
218	*Equisetum laevigatum*	Smooth horsetail			16
136	*Equisetum variegatum*	Variegated horsetail		83.4	
216	--	Wild flower			5.9
Sixth site					
47	*Picea sp.*	Spruce	5.1		5.2
48/141/220	*Lythrum salicaria*	Purple loosestrife	1.5	9.3	18.7
49	*Juniperus communis*	Common juniper	4.1		
50/137/219	*Equisetum laevigatum*	Smooth horsetail	27.6	53.6	59
51	*Aster lanceolatus*	Panicle aster	5.8		
52	*Populus balsamifera*	Poplar balsam	1.1		

(Table Continued)

Table 4.1 (continued)

Plant ID (DL-#)	Latin name	Common name	Collected May 22, 03	Collected July 17, 03	Collected Sept. 18, 03
			Total As ($\mu g \cdot g^{-1}$)		
Sixth site (contd.)					
53	*Betula papyrifera*	Paper birch	0.9		
54	*Vicia sp.*	Vetch	3.1		
55/143/225	*Aster lanceolatus*	Panicle aster	6.7	14.9	10
56/138/221	*Equisetum arvense*	Field horsetail	90.1	58.5	26.3
139	*Ambrosia artemisiifolia*	Common ragweed		8.9	
140	*Puccinellia sp.*	Alkali grass		5.2	
142	*Lotus corniculata*	Bird's foot trefoil		3.5	
144/226	*Vicia cracca*	Blue Vetch		0.6	1.6
145/214	*Solidago canadensis*	Canada goldenrod		0.7	6.9
224	*Solidago canadensis*	Canada goldenrod			9.3
146	*Equisetum variegatum*	Variegated horsetail		425 ± 6	
223	*Pteridium aquilinum*	Bracken fern			2.3

[1] Plant 207b was collected on 31 August, 2004

Photo 4.5. An area near plant sampling Sites 1, 2, and 3 north of the Young's Creek and south of the arsenic tailing area is shown.

Photo 4.6. A general view of the arsenic sampling Sites 4, 5 and 6 near the industrial area and Deloro village is shown.

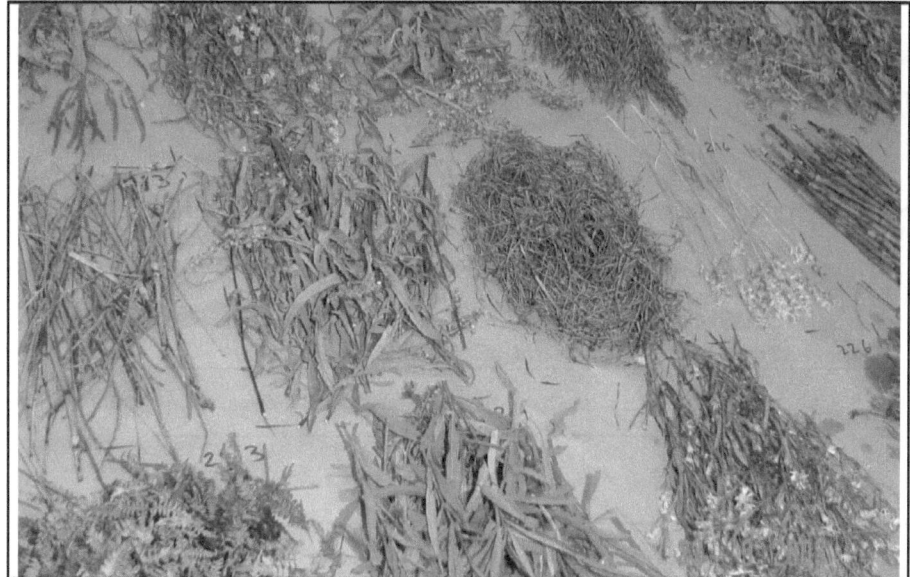

Photo 4.7. A view of the washed plants lay on the laboratory table for drying.

Photo 4.8. A photo of the author of the thesis 'facing arsenic without fear' is shown near the netted arsenic tailing pond in the background.

4.4.1. Sampling Variations of Arsenic Concentration in Plants and Soils

The soil arsenic concentrations of the sites along with the means, medians and ranges of arsenic concentrations in plants from the respective sites are reported in Table 4.2. The soil arsenic content varied from 424-6900 $\mu g.g^{-1}$. Such variation of contaminant

Table 4.2. Soil arsenic concentrations and mean, median and ranges of plant arsenic concentrations in samples collected from six sites of Deloro gold mine area.

	Site 1	Site 2	Site 3	Site 4	Site 5	Site 6
Soil As Concentration, $\mu g.g^{-1}$	424	807	6900	530	6060	2100
Plant As Concentration, $\mu g.g^{-1}$						
Mean	42.7	5.9	21.9	1.2	30.5	29.8
Median	5.1	3.8	4.35	1.2	7.7	6.7
Range	0.6 - 532	0.5 - 19.7	0.4 - 157	0.8 - 1.5	0.4 - 210	0.6 - 425
No. of Plants (n =)	34	17	14	2	28	29

concentration in the soils of an impacted area is expected given the heterogeneous metal content of mine wastes[86] and the uncontrolled movement of these wastes over the site for many years. Wide ranges of arsenic in plants were observed in sites 1, 3, 5, and 6. In this study, a direct correlation between soil and plant arsenic concentrations was not observed. This was likely due to the diversity of the plant species and their uptake of arsenic from soils as well as the soil conditions in different areas. The results of a study concerning the transfer of arsenic from soil to plant using controlled soil and soil slag mixers have indicated that the ability of plants to uptake arsenic and other metals may vary depending on soil conditions.[143] Further studies confirm that arsenic uptake by plants has been dependent on plant species[144] and soil acidity.[145,146] Because of the diversity in the soil arsenic concentrations and the plant species over a large site, a direct comparison of the arsenic content in soil and plant cannot be made.

Arsenic in Plants: Extraction and Speciation

In the determination of arsenic in plants in this study, a number of plants of the same species collected from one spot (of about 1 sq. m area) were mixed together and prepared for all sampling and analysis. This gave average concentrations of the arsenic species in the plant from the spot. However, variation of arsenic concentration within the plants of same species from a small area was not known. In order to assess the variation of arsenic concentration within one species of plant and within soil samples from a spot, plant and soil samples were collected from 1-square meter plots and analyzed individually. Testing for any correlation between soil and plant arsenic concentrations was conducted. The relationship of arsenic with other essential and nonessential elements based on the determination of multi-elements in plants will be discussed in Chapter 5.

Three different plant species from three different areas were sampled. Five samples of each species and five soil samples from each area were collected. All soil and plant samples were prepared and analyzed individually. Details of sample collection and preparation are provided in Chapter 2. The total arsenic in the plant and soil samples were determined following the standard ASU methods for arsenic in plants and soils also described in Chapter 2.

The results of soil and plant sample analyses are reported in Table 4.3. The field horsetails (PL-3) were mature plants and close to their decay state at the end of the growing season. The average (n = 5) arsenic concentrations determined from five independent determinations in plants PL-1, PL-2 and PL-3 were 22.6 ± 5.8, 9.1 ± 1.3 and 7.9 ± 2.7 $\mu g.g^{-1}$, respectively. The average (n = 5) arsenic concentrations ($\mu g.g^{-1}$) in the corresponding soil samples S-1, S-2 and S-3 were 3300 ± 636, 4140 ± 2650 and 683 ± 612, respectively.

Table 4.3. Determination of the variation of arsenic concentrations in plants and soils collected from one-meter square plots of Deloro gold mine area.

Soil samples				Plant samples			
Sample ID	Total As ($\mu g.g^{-1}$)	Ave. conc. ± 1 s.d. ($\mu g.g^{-1}$)	RSD	Sample ID	Total As ($\mu g.g^{-1}$)	Ave. conc. ± 1 s.d. ($\mu g.g^{-1}$)	RSD
S-1.1	3440	3300 ± 636	19	PL-1.1	28.7	22.6 ± 5.8	26
S-1.2	4115			PL-1.2	14.3		
S-1.3	2360			PL-1.3	27.0		
S-1.4	3448			PL-1.4	19.8		
S-1.5	3126			PL-1.5	23.4		
S-2.1	2861	4140 ± 2650	64	PL-2.1	6.9	9.1 ± 1.3	14
S-2.2	8836			PL-2.2	9.0		
S-2.3	3416			PL-2.3	10.2		
S-2.4	3026			PL-2.4	9.9		
S-2.5	2546			PL-2.5	9.6		
S-3.1	388	683 ± 612	90	PL-3.1	11.5	7.9 ± 2.7	30
S-3.2	203			PL-3.2	8.7		
S-3.3	1114			PL-3.3	6.2		
S-3.4	170			PL-3.4	4.5		
S-3.5	1539			PL-3.5	8.9		

PL-1 = Smooth horsetail (*Equisatum laevigatum*); PL-2 = Canada goldenrod (*Solidago Canadensis*); PL-3 = Field Horsetail (*Equisatum arvense*).

For each plant sample, a soil sample was collected close to its roots. Little or no association between plant and soil arsenic concentrations was observed. This indicates that in the presence of sufficient arsenic, uptake of arsenic is limited by the plant species rather than the amount of arsenic in the soil near their roots. However, a soil sample close to the root of the plant may not necessarily be representative of the surrounding soil the plant is exposed to depending on the size of the root system of the plant. Study of elemental composition in several plant species has shown that plants actively accumulate or eliminate some elements although grown on the same soil under the same environmental conditions.[147]

Soil and plant samples, S-1/PL-1, S-2/PL-2 and S-3/PL-3, correspond to those from Site 6, Site 5 and Site 2, respectively, and are reported in Table 4.2. The total arsenic in the soil and plant samples from these sites agreed well. Arsenic in Canada goldenrods (PL-2) determined in Table 4.3 compared favorably with those in Table 4.1 collected in September. The RSD values of total average arsenic concentrations (n = 5) within the plant species are in the statistical tolerance range of about 30%.

4.5. ARSENIC IN DIFFERENT PLANT SPECIES

Seven plant species collected over the period May-September of 2003 from Deloro are summarized in Table 4.4. Among the species, it can be seen that the total arsenic content varied widely. The highest arsenic accumulation was observed in a field horsetail (DL-15, 532 µg.g^{-1}) that grew in the water body with its roots under water; arsenic concentrations within this group of 12 plants varied with an overall mean of 159 µg.g^{-1}. For the most part, the horsetails, which are members of the fern family, accumulated more arsenic (overall average: 127 µg.g^{-1}) than any other species reported in the table (Table 4.4). Only another fern (sensitive fern) accumulated higher arsenic (26.4 µg.g^{-1} mean) compared to other species but still much less than the horsetails.

In a field study searching for plant markers toward gold deposits,[120] field horsetails were found to contain high amounts of arsenic compared to the amounts of Au and Sb near gold mineralization, and an arsenic concentration as high as 738 µg.g^{-1} in one horsetail (*E. fluviatile L.*) was reported. The presence of high arsenic in the field horsetails from an abandoned goldmine in Goldenville, Nova Scotia, Canada was also reported.[121] The three plants of the *Equisetum* family, *E. fluviatile*, *E. hyemale* and *E. sylvaticum*, accumulated 371 µg.g^{-1}, 365 µg.g^{-1} and 418 µg.g^{-1} arsenic, respectively, on the tailing field at Goldenville and their potential for phyto-remediation of arsenic has been observed. These fern allies (i.e., the *Equisetum sp.*), however, did not show the hyper-accumulation as has been reported in Chinese brake fern (*Pteris vittata L.*).[87,88] However, in contrast to Chinese brake ferns, horsetails grow and thrive in cold climates and, therefore, may be used as phyto-remediators in Canada and other cold weather places.

Table 4.4. Total arsenic in various plant species collected from different sites of Deloro gold mine area.

Common Name[a]	Plant ID Number (DL-#)	Site of Collection	Total As in Plants, µg.g^{-1}	Mean As Conc., µg.g^{-1}
Sensitive fern	6, 104, 201	1	8.6, 22.1, 48.4	26.4
Canada goldenrod	18, 119, 206	2	3.8, 5.8, 1.1	5.1
Canada goldenrod	129	5	8.2	
Canada goldenrod	145, 214, 224	6	0.7, 6.9, 9.3	
Field horsetail	14, 15, 101	1	295, 532, 410	159
Field horsetail	118, 204	2	19.7, 19.4	
Field horsetail	60, 124, 207	3	24.9, 59.5, 157	
Field horsetail	130	5	210	
Field horsetail	56, 138, 221	6	90.1, 58.5, 26.3	
Smooth horsetail	135, 212, 213, 218	5	36.8, 42, 10.5, 16	35.1
Smooth horsetail	50, 137, 219	6	27.6, 53.6, 59	
Variegated horsetail	136	5	83.4	254
Variegated horsetail	146	6	425	(127)[b]
Panicle aster	20, 121, 205	2	1.9, 9.6, 2.9	7.6
Panicle aster	134	5	8.9	
Panicle aster	51, 55, 143, 225	6	5.8, 6.7, 14.9, 10	
Purple loosestrife	109	1	11.2	
Purple loosestrife	41, 128, 209	5	2.3, 18.6, 20.3	11.7
Purple loosestrife	48, 141, 220	6	1.5, 9.3, 18.7	

[a]Scientific names of the plants are given in Table 4.1. [b]Overall average As concentration of all horsetails.

The *Onoclea sensibilis* (sensitive fern) sampled from site 1 (DL-6/DL-104/DL-201) showed a gradual increase in arsenic content over the months to a maximum of 48.4 µg.g^{-1} compared to 23.5 µg.g^{-1} found in the same species in another study.[148] In the present study, *Solidago Canadensis* (Canada goldenrod) was found to accumulate the least amount of arsenic among the plants presented in Table 4.4; the arsenic concentrations within this group of plants varied from 0.7-9.3 µg.g^{-1} with 5.1 µg.g^{-1} mean concentration. The total arsenic in *Typha sp.* (cattail) found in site 1 (Table 4.1) was determined to be 1.4 µg.g^{-1} in May, and those cattails grown on site 5 were found to contain 0.5 µg.g^{-1} and 0.9 µg.g^{-1} in May and July, respectively. *Typha sp.* has been known to be less arsenic accumulating but thrives well in the high arsenic soil.[149] Except for the *Equisetum* species and *Onoclea sensibilis* on sites

1 and 5, the other plants in this study did not accumulate arsenic much higher than 10 µg.g^{-1}. The *Lythrum salicaria* (purple loosestrife) absorbed arsenic in the range of 10-20 µg.g^{-1} in all soils. The *Aster lanaceolatus* (panicle aster) sampled from sites 2, 5 and 6 showed a maximum accumulation of 14.9 µg.g^{-1}. These results showed that the absorbance and accumulation of arsenic by the plants was species dependent.

4.6. VARIATION IN ARSENIC CONCENTRATION IN PLANTS DURING THE GROWING SEASON

The arsenic results of plant species collected from the same areas on the three expeditions to Deloro in May, July and September of 2003 are reported in Table 4.5. Six

Table 4.5. Seasonal variation of arsenic concentrations in plants collected at different times of the season. Arsenic in same plant species collected from same site are compared. For a visual inspection of the changes see Figure 4.1.

Plant ID (DL-#)	Site	Common plant name[a]	May 22, 2003 (µg.g^{-1})	July 17, 2003 (µg.g^{-1})	Sept. 18, 2003 (µg.g^{-1})
5/106/200	1	Wild strawberry	2.35 ± 0.01[b]	10.7	10.4
6/104/201	1	Sensitive fern	8.6	22.1 ± 1.0[b]	48.4
18/119/206	2	Canada goldenrod	3.8	5.8	1.1
20/121/205	2	Panicle aster	1.9	9.6	2.9
21/118/204	2	Field horsetail	15	19.7 ± 0.2[b]	19.4
60/124/207	3	Field horsetail	24.9 ± 0.6[b]	59.5 ± 3.0[b]	157 ± 4[b]
41/128/209	5	Purple loosestrife	2.3	18.6	20.3
48/141/220	6	Purple loosestrife	1.5	9.3	18.7
50/137/219	6	Smooth horsetail	27.6	53.6	59
55/143/225	6	Panicle aster	6.7	14.9	10
Averages	--	--	9.5 ± 9.8	22.4 ± 18.8	34.7 ± 46.9

[a]Scientific names are given in Table 4.1. [b]Errors are ± ave. deviation from two independent determinations.

of the plant species (wild strawberry in site 1, Canada goldenrod in site 2, panicle asters in sites 2 and 6, purple loosestrife in site 5, and smooth horsetail in site 6) accumulated arsenic until the middle of the season. Three plants (sensitive fern in site 1, field horsetail in site 3, and purple loosestrife in site 6), on the other hand, continued accumulating until at the end of the season. A significant increase (by paired t-test for means), from 9.5 ± 9.8 µg.g^{-1} to 22.4 ± 18.8 µg.g^{-1}, on the average amount of arsenic from May to July plants was observed. July plants were vigorous at the height of their biological activity and growth declined toward the end of the season. The stems and leaves of many plants became dried, pigmented and in some cases, insect consumed or infested at the end of the growing season.

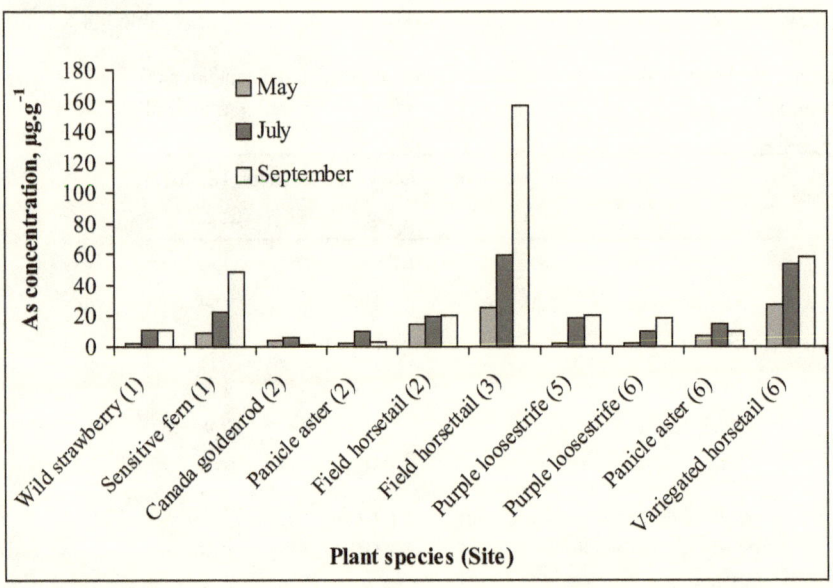

Figure 4.1. Bar graphs show seasonal changes in arsenic concentrations in a number of plant species. Sampling sites are shown in parentheses. See Table 4.5 for details.

The pattern of continued arsenic uptake became less clear in the September plants although more average arsenic was found in these plants than in the July plants. A graphical representation of the seasonal change in the arsenic concentrations in plants of Deloro is shown in Figure 4.1. As already mentioned above, the plants were not sampled from identical locations; the six sampling sites were general locations and the soil arsenic content varied significantly in the area. These results, however, indicate a general trend of increased concentration over the season with horsetails, sensitive fern and purple loosestrifes continuing to uptake arsenic during the season.

4.7. ANALYSIS OF ARSENIC IN DIFFERENT PARTS OF PLANTS

The arsenic concentrations determined in the different parts of plants are reported in Table 4.6. The total arsenic content in the different parts of plants was determined

Table 4.6. Determination of arsenic in different parts of plants. For plant scientific names refer to Table 4.1.

Common name	Plant ID (DL-#)	Levels of arsenic in plant parts		
		Stem, µg.g^{-1}	Leaves, µg.g^{-1}	Flowers, µg.g^{-1}
Poplar	30	0.92	1.1	NA
Willow	31	0.33	1.1	NA
White pine	32	nd	6.3	NA
Purple loosestrife	128	4.89	18.6	NA
Purple loosestrife	209	5.11	20.3	1.05
Canada goldenrod	129	6.53	8.2 ± 0.4^a	NA
Canada goldenrod	208	6.68	9.3 ± 0.1^a	0.91
Canada goldenrod	214	4.44	6.9	0.57

^{a}Mean ± ave. deviation from two independent determinations, nd = not detected, and NA = flowers not available at the time of season.

following the same procedure as was used for the ground plant samples. Poplar showed similar concentrations in stem (0.92 µg.g^{-1}) and leaves (1.1 µg.g^{-1}), but willow leaves and needles of white pine contained more arsenic than the stems did. The leaves of other plants contained more arsenic than the stems or flowers did. Higher amounts of arsenic in matured leaves and common leafy vegetables were also observed in other studies.[84,150] The leaves, apart from the roots, act as the powerhouse of plants containing many enzymes and proteins, and as the respiratory organ of the plant, are expected to accrue more arsenic due to binding to the thiolate groups of the plant peptides, and for complexation with the phytochelatins.[151]

The presence of arsenic in the leaves and edible stems above ground has an important implication for the health of humans and wildlife. The results from the laboratory and field evaluations have suggested correlations between the arsenic concentration in soil and elevated risks to humans, vegetation, and wildlife.[49,50,152] Serious concerns regarding health risks due to elevated arsenic in the environment were expressed in scientific reviews[19,85] and supported by the toxicological and pharmacological tests.[28,31,153] At Deloro, however, risks to wildlife have not been a major concern according to the Draft Cleanup Plan Summary for Deloro authored by the Ontario Ministry of Environment.[57] In our trips to the site, we observed deer foot-steps on the arsenic ground near the industrial area north of water treatment tailing pond and on soil tracks in the woods near the Young' Creek area. A food chain study on deer mice was conducted in Yellowknife, NWT.[154] To our best knowledge, there has been no arsenic food chain study conducted in the Deloro area. This study is important because animals such as deer and beavers live on plants grown on the arsenic contaminated soils and water bodies of the area. The long-term effects on the bird population of the area that feeds on the small animals and insects that in turn feed on the plants of the area are not known. However, preliminary studies have shown elevated levels of As in insects sampled from the area (personal communication with Dr. John S. Poland).

4.8. EXTRACTION AND SPECIATION OF ARSENIC BY SEQUENTIAL METHOD The method of arsenic extraction by 1:1 water-methanol and 0.1 M-HCl sequentially was developed and optimized in the present study, and applied to the terrestrial plants from Yellowknife in NWT, Canada. The development and application of the sequential method has been described in Chapter 3. After the initial investigation with a number of plants sampled from Deloro in November 2001, more plant and soil samples were collected in May, July and September of 2003 from the area. Plants were extracted for arsenic by the sequential method (0.25 g sample, 3.0 mL DDW regeneration of 1:1 water-methanol extracts) developed by the current investigation.

Table 4.7. Sequential extraction and speciation of plants collected in May 2003 from Deloro.

Plant ID (DL-#)[a]	Speciation by HPLC-HGAAS[b] ($\mu g \cdot g^{-1}$)		Sum of species ($\mu g \cdot g^{-1}$)	Total As in plants ($\mu g \cdot g^{-1}$)	% EE
	As(III)	As(V)			
6	nd	6.60	6.60	8.60	77
50	12.8	14.9	27.7	27.6	100
56	51.4	43.9	95.3	90.1	106
60	11.1	10.6	21.8	24.9 ± 0.6[c]	88

[a] Plants are listed in Table 4.1. [b] Water-alcohol and acid extracts were speciated and results were combined. Organoarsenic species were detected in water-alcohol extracts only. No DMA in these plants but 0.10 µg.g-1 MA was detected in DL-60. [c] Mean ± ave. deviation from two independent analyses. nd = species not detected.

The total arsenic of the plants was determined (Table 4.1) and those samples that contained about 5 µg.g⁻¹ arsenic or more were extracted by the sequential method. The speciation of arsenic in the 1:1 water-methanol and 0.1 M-HCl extracts was performed by HPLC-HGAAS. Arsenic species in the two extracts were determined independently. The results of speciation analysis were combined by summing the concentrations of the individual arsenic species and are reported in Table 4.7 for the May plants, Table 4.8 for July plants and Table 4.9 for September plants.

The total arsenic in the plants and the percent extraction or extraction efficiency (EE) of sequential method is also reported in the tables. Additional plants from Deloro were extracted and speciated. These low arsenic plants were speciated for the organoarsenic species in 1:1 water-methanol extracts only and total arsenic in the HCl extracts was determined by ICPAES. The results of the latter extraction and speciation are presented in Table 4.10.

Table 4.8. Sequential extraction and speciation of plants collected in July 2003 from Deloro. Mean ± ave. deviation from two independent analyses of a number of plants are reported for QA.

Plant ID (DL-#)[a]	Speciation by HPLC-HGAAS[b] ($\mu g \cdot g^{-1}$)			Species sum, ($\mu g \cdot g^{-1}$)	Total As, ($\mu g \cdot g^{-1}$)	% EE
	As(III)	DMA	As(V)			
101	166 ± 6		181 ± 12	347 ± 18	410	85
103	2.6		3.7	6.2	11.2	56
104	5.1 ± 0.1		7.3 ± 0.1	12.4 ± 0.2	22.1 ± 1.0	56
105	0.8	0.40	2.1	3.3	8.4	40
106	2.5		2.7	5.2	10.7	49
108	2.0	0.13	3.4	5.6	11.3	50
109	2.2		0.2	2.41	11.2	21
110	1.9		8.3	10.2	13.2	77
115	0.6		5.8	6.3	9.1	70
118	17.5		1.4	18.9	19.7 ± 0.2	104
121	5.5 ± 0.3		1.21 ± 0.04	6.7 ± 0.3	9.6	70
124	36.4		12.7	49.1	59.5 ± 3.0	83
128	2.9		0.1	3.06	18.6	16
129		0.23	1.3	1.6	8.2 ± 0.4	20
130	149 ± 5		22.1 ± 2.2	171 ± 7	210	81
133	2.9		0.8	3.7	7.2	51
134	0.2		3.2	3.4	8.9	38
135	19.3		17.6	36.9	36.8 ± 0.6	100
136	41.9 ± 1.4		47.9 ± 4.2	89.8 ± 5.6	83.4	103
137	22.1		28.1	50.2	53.6	94
138	21.3		31.1	52.4	58.5	90
141	1.2 ± 0.1	0.10 ± 0.01	0.33 ± 0.05	1.6 ± 0.2	9.3	18
143	4.6		5.6	10.2	14.9	69
146	203		221.7	425	425 ± 6	100

[a]Plants are listed in details in Table 4.1. [b]Water-alcohol and acid extracts were speciated and results were combined. Organoarsenic species were detected in water-alcohol extracts only. 0.10 $\mu g \cdot g^{-1}$ MA was detected only in DL-124. Void spaces represent species not detected.

Table 4.9. Sequential extraction and speciation of plants collected in September 2003 from Deloro.

Plant ID (DL-#)[a]	Speciation by HPLC-HGAAS[b] ($\mu g \cdot g^{-1}$)				Species sum, $\mu g \cdot g^{-1}$	Total As, $\mu g \cdot g^{-1}$	% EE
	As(III)	DMA	MA	As(V)			
200	2.19	0.10		3.44	5.72	10.4	55
201	9.63			9.75	19.4	48.4	40
204	2.68		0.13	17.0	19.8	19.4	102
205	0.27	0.11			0.374	2.9	13
207[c]	13.6 ± 0.8		4.59 ± 0.04	116 ± 1	134 ± 2	157 ± 4	85
209	3.32	0.28		1.15	4.74	20.3	23
219	20.9		0.14	36.0	57.0	59.0	97
220	2.85	0.21		0.533	3.59	18.7	19
221	5.63		2.2	12.7	20.5	26.3	78
225	4.90			1.6	6.5	10.0	65

[a]Plants are listed in Table 4.1. [b]Water-alcohol and acid extracts were speciated and results were combined. Organoarsenic species were detected in alcohol extracts only. Void spaces represent species not detected. [c]Mean ± ave. deviation from two independent determinations.

Table 4.10. Additional plants from Deloro sampled in May, July and September 2003 were extracted and speciated for organoarsenic species. Mixed spike recovery experiment was carried out along with these plants and results are reported in this table and discussed in Section 3.7.1.

Plant ID (DL-#)[a]	Total As ($\mu g \cdot g^{-1}$)	% EE	Arsenic species in 1:1 Water-methanol extracts ($\mu g \cdot g^{-1}$)			
			As(III)	DMA	MA	As(V)
12	6.4	27	0.30	nd	nd	nd
18	3.8	50	nd	nd	nd	0.93
32	6.3	65	0.47	nd	nd	3.1
102	3.6	67	1.1	nd	nd	0.78
116	3.9	72	0.36	nd	nd	1.9
119	5.8	36	nd	0.23	nd	1.1
125	5.3	50	1.6	nd	nd	1.1
126	3.4	68	nd	nd	nd	1.9
206	1.1	92	nd	nd	nd	0.51
Spikes[b]	0.80	91	0.18	0.15	0.21	0.19

[a]Plants are listed in Table 4.1. [b]Concentration of each spike was 0.20 $\mu g \cdot g^{-1}$. nd = not detected.

The speciation results showed that the inorganic arsenic, As(III) and As(V), was the major constituent in all plants and small amount of organoarsenic species, MA and DMA, was detected in a number of plants. It is interesting to note that MA was detected only in the horsetails, mainly the field horsetail, and DMA was detected in a number of other plant species. Detailed discussions on arsenic speciation are given in the subsequent sections (4.8.1 and 4.8.2). An inspection of Tables 4.7, 4.8, 4.9 and 4.10 reveals that extraction efficiencies vary widely for the plant species from as low as about 13% to as high as 104%. Although most arsenic from the horsetail family could be extracted, purple loosestrifes, on the other hand, afforded the least amount of arsenic from their matrices ranging 16-23% with a median at 19%.

The sequential method of arsenic extraction adopted the traditional method of water-alcohol extraction of plants plus the dilute hydrochloric acid extraction aided by sonication in both solvents in order to enhance the extraction efficiency of the process. One of the important advantages of the method was found to be the ability of determining the trace organic species in plants containing high amount of inorganic arsenic. This was achieved by separating the organic species in a medium that suppressed the extraction of the inorganic species. Further, the method of removal of the organic solvent and regeneration in a much smaller volume of water, developed by the current study, enhanced the detectability of the trace organoarsenic species by the HPLC-HGAAS technique.

The speciation results of 1:1 water-methanol and 0.1 M-HCl solvents are reported separately in Table 4.11. Percent extraction efficiencies (EE) of 1:1 water-methanol and the sequential method are reported. A comparison of the overall extraction efficiency of the sequential method with the overall extraction efficiencies of 1:1 water-methanol was made. The overall total amount of arsenic in the plants was 2050 $\mu g.g^{-1}$ and the overall extracted total amount was 1750 $\mu g.g^{-1}$. The 1:1 water-methanol and 0.1 M-HCl extracted 896 $\mu g.g^{-1}$ and 843 $\mu g.g^{-1}$, respectively, in sequential extractions. The results of overall EE calculations using the speciation data of Table 4.11 are reported in Table 4.12.

Thus, water-methanol, the traditional solvent for extraction, extracted 44% of the overall total arsenic and 51% of the overall total extracted arsenic. Had the extraction been conducted with the traditional water-methanol solvent alone, an overall extraction efficiency of only 44% would have been achieved for these plants. In contrast, almost 100% more arsenic was extracted by the 0.1 M-HCl sequential extraction of the plants giving an overall improvement of 85% extraction efficiency. An improved range of 17-106% for sequential method was observed compared to 1-89% for the water-methanol extraction.

Table 4.11. The speciation results of As in 1:1 water-methanol and 0.1 M-HCl extracts of plants from Deloro are reported separately. EE's of two method are compared. No DMA or MA was detected in HCl medium. Mean ± ave. deviation from duplicate determinations are provided to show QA. Scientific and common English names of the plants are given in Table 4.1. Void spaces indicate no species detected.

Plant ID (DL-#)	1:1 Water-methanol extraction						0.1 M-HCl Extraction				% EE, Sequential method
	As(III), $\mu g \cdot g^{-1}$	DMA, $\mu g \cdot g^{-1}$	MA, $\mu g \cdot g^{-1}$	As(V), $\mu g \cdot g^{-1}$	Species sum, $\mu g \cdot g^{-1}$	% EE	As(III), $\mu g \cdot g^{-1}$	As(V), $\mu g \cdot g^{-1}$	Species sum, $\mu g \cdot g^{-1}$	
6				4.32	4.32	50		2.27	2.27	77
50	4.3			14.5	18.8	68	8.4	0.45	8.9	100
56	47.2			32.9	80.1	89	4.1	11.0	15.1	106
60	10.7		<0.10	7.7	18.5	74	0.46	2.9	3.3	88
101	15.8 ± 11.2			110 ± 12	125 ± 1	30	150 ± 6	71.60 ± 0.04	222 ± 6	85
103	0.27			3.7	3.9	35	2.3		2.3	56
104	0.17 ± 0.02			5.9 ± 0.1	6.0 ± 0.1	27	5.0 ± 0.1	1.5 ± 0.3	6.4 ± 0.3	56
105		0.40		2.1	2.5	30	0.82		0.82	40
106				2.7	2.7	25	2.5		2.5	49
108	0.22	0.13		3.4	3.8	33	1.8		1.8	50
109				0.18	0.18	1.6	2.2		2.2	21
110	0.33			4.6	4.9	37	1.6	3.7	5.3	77
115				3.9	3.9	43	0.55	1.9	2.4	70
118	13.4			<0.1	13.5	68	4.1	1.3	5.4	104
121	4.5 ± 0.2			1.21 ± 0.04	5.7 ± 0.2	59	1.0 ± 0.1		1.0 ± 0.1	70
124	21.6		<0.10	5.4	27.1	46	14.7	7.3	22.0	83
128				0.12	0.15	0.79	2.9		2.9	16
129		0.23		1.3	1.6	19				20

(Table continued)

Table 4.11 (continued).

Plant ID (DL-#)	1:1 Water-methanol extraction						0.1 M-HCl Extraction				% EE, Sequential method
	As(III), µg·g^{-1}	DMA, µg·g^{-1}	MA, µg·g^{-1}	As(V), µg·g^{-1}	Species sum, µg·g^{-1}	% EE	As(III), µg·g^{-1}	As(V), µg·g^{-1}	Species sum, µg·g^{-1}	
130	80.8 ± 6.2			3.9 ± 1.7	79.1	38	68.6 ± 1.4	18.2 ± 3.9	86.8 ± 5.3	81
133	2.9			0.78	3.7	51				51
134	0.23			2.6	2.9	32		0.54		38
135	2.9			16.3	19.2	52	16.4	1.3	17.7	100
136	5.3 ± 1.2			41.2 ± 3.4	46.5 ± 2.2	56	36.6 ± 2.6	6.7 ± 0.7	43.3 ± 3.3	103
137	2.7			17.9	20.6	38	19.5	10.1	29.6	94
138	6.5			20.9	27.4	47	14.8	10.2	25.0	90
141	0.122 ± 0.002	0.104 ± 0.001		0.33 ± 0.05	0.55 ± 0.06	5.9	1.1 ± 0.1		1.1 ± 0.1	18
143	4.5			5.6	10.1	68	0.15		0.15	69
146	59.3			191	251	59	144	30.5	174	100
200		<0.10		2.5	2.6	25	2.2	0.93	3.1	55
201	0.18			7.6	7.8	16	9.5	2.1	11.6	40
204	0.96		0.13	5.3	6.4	33	1.7	11.7	13.4	102
205	0.27	0.11			0.37	13				13
207	6.8 ± 0.5		4.59 ± 0.03	38.2 ± 2.7	49.6 ± 2.2	32	6.7 ± 0.3	77.7 ± 3.4	84.4 ± 3.6	85
209		0.28		0.50	0.78	3.8	3.3	0.64	4.0	23
219	8.7		0.14	21.2	30.0	51	12.2	14.7	26.9	97
220		0.21		0.46	0.67	3.6	2.9	<0.1	2.9	19
221	3.8		2.2	3.8	9.7	37	1.9	9.0	10.8	78
225	4.1			1.0	5.1	51	0.80	0.56	1.4	65

Table 4.12. Comparison of the overall extraction efficiency of the 1:1 water-methanol traditional method and sequential method.

Method	% EE of total As in plants	% EE of extracted As from plants	Range of EE
1:1 Water-methanol extraction	44	51	1 - 89%
Sequential extraction	85	100	17 - 106%

The published works of arsenic extractions using water-alcohol media showed similar results. Koch et al. showed that in general only about 50% of the arsenic could be extracted from plants by using 1:1 water-methanol and sonication methods.[15,16,92] Using 1:9 water-methanol only 2.5-12% arsenic has been extracted from plants grown on top of an ore vein[93] and 6.3-16.1% from plants of Moira watershed (Canada).[102] Using 1:2 water-methanol solvent, 73% arsenic has been extracted from a certified reference material (GBW 82301-peach leaves).[95] The predominance of As(V) extracted in the water-methanol solvents has been observed also in these studies (see Table 4.13 in Section 4.8.1). Since our experiments using 0.1 M-HCl as single extractant (described in Chapter 5) showed a complete or nearly complete extraction of the extractable inorganic species by the acid medium, a preferential extraction did not occur in this medium.

The sequential extraction method was developed to attain high EE and at the same time maintain the integrity of the species. Our studies detailed in Chapter 3 and Chapter 4 have shown that using the sequential extraction method the preservation of species could be maintained and much higher EE, compared to the literature values, could be achieved. The presence of organic species in terrestrial plants is found to be very low in concentration and since MA and DMA were the most common species in plants and most toxic after arsenite and arsenate, we looked for these two organoarsenic species in the Deloro plants. Subsequent study with HPLC-ICPMS confirmed MA and

Arsenic in Plants: Extraction and Speciation

DMA in the plants and did not detect any other species in the extracts (HPLC-ICPMS results will be discussed in Chapter 5). The inorganic arsenic, As(III) and As(V), in most of the plants comprised more than 99% of the total arsenic extracted.

4.8.1. Extraction of As(III) and As(V) by Sequential Method: Their Influence on the Extraction Efficiencies

The question, whether arsenite (As(III)) or arsenate (As(V)) are present, has been considered all along in the history of speciation of arsenic in the environment.[6,11,27] The knowledge of these species is important because of their relatively high toxicity. The influence of the species in the absorption and assimilation of arsenic by plants has been investigated.[144] However, the effect of species on the extraction efficiency of arsenic from plants has not been studied. A general trend in the extraction of species in the 1:1 water-methanol and 0.1 M-HCl media is presented and discussed here. Note that the HCl extraction is performed on the residue after it has been extracted by 1:1 water-methanol solvent.

Table 4.13. Ratios of arsenate to arsenite in 1:1 water-methanol, 0.1M-HCl and overall ratio in the extracts of plants calculated from data reported in Table 4.11.

Sample description	In 1:1 Water-methanol	In 0.1 M-HCl	In overall extracts of plants
	As(V)/As(III)	As(V)/As(III)	As(V)/As(III)
All plants excluding horsetails (22)	3.06	0.35	1.18
All horsetails (16)	1.82	0.57	1.03
All plants (38)	1.89	0.55	1.04

The ratios of arsenate to arsenite (As(V)/As(III)) extracted from plants by the two solvents are listed in Table 4.13. The overall extracted As(V)/As(III) ratios in 'all plants excluding horsetails,' 'all horsetails' and 'all plants' are 1.18, 1.03 and 1.04, respectively, and those in 1:1 water-methanol are 3.06, 1.82, and 1.89, respectively. The enhanced As(V)/As(III) ratios in 1:1 water-methanol showed that this medium extracted more arsenate than arsenite. The remaining arsenic was extracted by 0.1 M-HCl and in this second extraction As(III) was the predominant species. The more arsenite in the HCl media may be indicative of the nature of As(III) chemically binding to the free or thiol sulfur in the plant matrix. HCl was possibly extracting lipid and/or protein bound arsenic in the plant matrix, too.[27]

The ratio of arsenate to arsenite (As(V)/As(III)) in plants and its relation to the total absorbed arsenic was investigated. The ratios As(V)/As(III) extracted by the sequential method from a number of plant species - field horsetails, panicle aster and purple loosestrife - were plotted as a function of both the total arsenic in plants and the extracted total arsenic from plants. The R^2 values from linear fits are shown for each graph and presented in Figure 4.2. A visual inspection as well as comparison of the R^2 values revealed that there was no correlation between the As(V) and As(III) distribution and the total arsenic absorption by plants either in the original plant samples or in the extracts. These findings showed that plants do not follow any common rule regarding the assimilation of the absorbed arsenic in their systems. Results of heterogeneous accumulation rate and metabolization by plants, as well as disproportionate inorganic arsenic species in plants have been reported for controlled substrate experiments.[144,155] However, a direct demonstration of an association between the As(V) to As(III) ratio and the total absorbed arsenic in plants naturally grown on arsenic impacted areas has not been reported.

On the other hand, the extraction efficiency of arsenic, which is of much interest to the researchers, appeared to have been affected by the ratio of the species present in the plants. The ratios of arsenite to arsenate (As(III)/As(V)) of twelve plant species have been compared with the %EE of arsenic from these plants. A graphical representation of the comparison is shown in Figure 4.3.

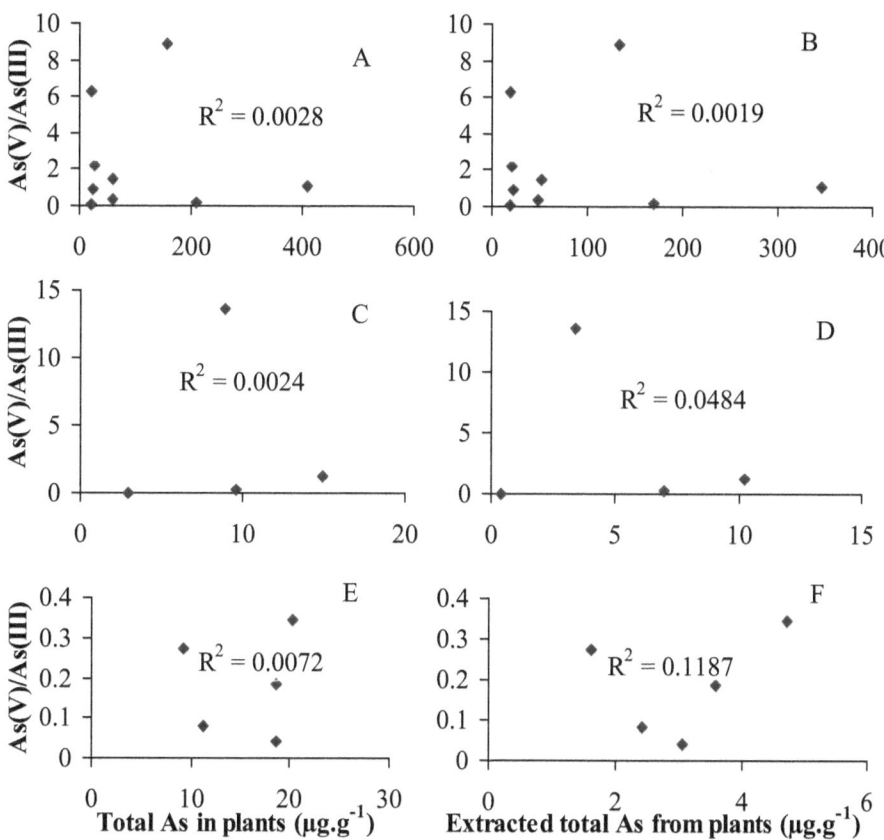

Figure 4.2. The ratios of arsenate to arsenite are plotted against the total arsenic in plants and the total arsenic extracted by sequential method. Values of R^2 from linear regressions are shown for each category. A and B: Field horsetails; C and D: Panicle aster; and E and F: Purple loosestrife.

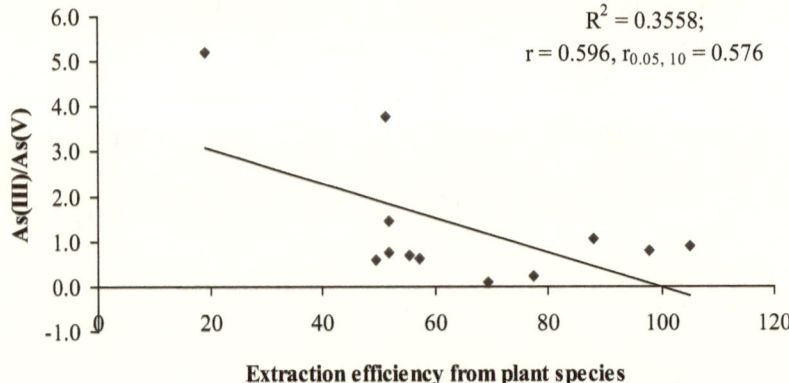

Figure 4.3. The ratios of arsenite to arsenate plotted against the arsenic extraction efficiencies of twelve plant species by the sequential method.

A significant correlation with a negative slope of linear regression line of the As(III)/As(V) ratio in plants to the efficiency of extraction (EE) of arsenic was observed. This showed that the presence of more As(III) in plants would make extraction more difficult causing lower EE. This is a significant observation in the understanding of extraction of arsenic from plants and, to the best of our knowledge, this has not been reported before. This finding is supported by the fact that in general water-alcohol extracts more arsenate than arsenite (Table 4.13). Arsenite is likely more difficult to extract due its chemical bonding with the plant matrix. It has been shown that plants metabolize arsenic by the complexation of arsenic with glutathione (GSH) and phytochelatins (PC, metal binding peptides derived from GSH) by forming As(III)-(GS)$_3$ and As-PC$_n$ complexes, respectively.[85,129] Hydrochloric acid may well break the As-S (arsenic-sulfur) bond in these complexes but that certainly makes its job difficult particularly when the arsenic is sequestered in those hard to reach plant vacuoles or other cellular systems.

4.8.2. Organoarsenic Compounds in Deloro Plants

As described in Section 1.2.1., the organoarsenic compounds are the minor arsenic species in plants with MA and DMA being the most commonly detected species. AB, TMAO, Tetra and arsenosugars

Arsenic in Plants: Extraction and Speciation

have been detected in lichens and mushrooms and in some plants as well.[11,26,92] In this study, MA was determined in the field horsetails, DL-207a, DL-207b and DL-221 (Table 4.9), to be 4.6 µg.g^{-1}, 8.0 µg.g^{-1} and 2.2 µg.g^{-1}, respectively. MA was also detected in field horsetails collected in May (DL-60, Table 4.7) and in September (DL-204); and in a smooth horsetail (DL-219), also from September collection, Table 4.9.

DMA was found in plants collected in both July and September. The blue vervain (DL-105) and Canada goldenrod (DL-129) collected in July contained 0.40 µg.g^{-1} and 0.23 µg.g^{-1} DMA, respectively. DMA was also detected in meadowsweet (DL-108) and purple loosestrife (DL-141), and the results are reported in Table 4.7. Amounts of DMA were found to be 0.28µg.g^{-1} and 0.21µg.g^{-1}, respectively, in purple loosestrife, DL-209 and DL-220 collected in September (Table 4.9).

Representative chromatograms of the sequential extraction of two Deloro plants are shown in Figure 4.4. The chromatograms depict extraction of the four arsenic species, As(III), DMA, MA and As(V), by the 1:1 water-methanol and 0.1 M-HCl sequential extractions. Both MA and DMA were not simultaneously detected in any of the plants in this study.

4.9. EXTRACTION OF ARSENIC: DEPENDENCE ON PLANT MATRIX

The dependence of extraction efficiency (EE) on the total arsenic content of plants has been reported earlier in Chapter 3 (Fig 3.6 and Fig. 3.7). The results showed that the efficiency of extraction declined as the total arsenic in plants increased. The earlier report was based on the EE of arsenic from different plant species. The dependence of EE on the total arsenic content of plants has been further studied using the same plant species thus minimizing the effect of difference in the matrices of plants on the relationship. The EE of the sequential method for the plant species were compared and the results are reported in the graphic forms in Figure 4.5.

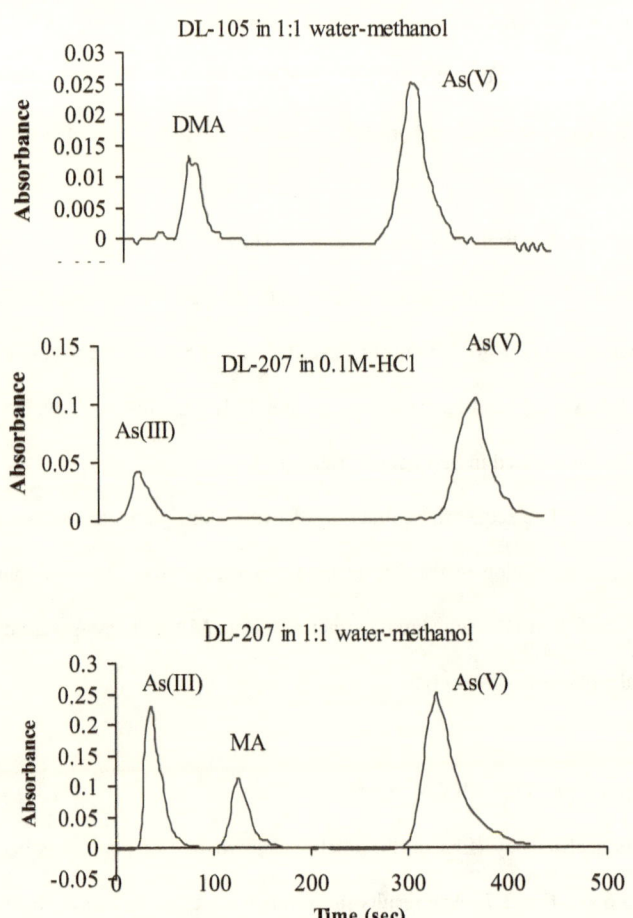

Figure 4.4. Representative HPLC-HGAAS chromatograms of Deloro plants extracted in 1:1 water-methanol and 0.1 M-HCl sequentially. HPLC was performed with an anion exchange column (Hamilton PRP-X100 250 x 4.6 mm column), with 20 mM ammonium phosphate, pH 6.0 at 1.5 mL.min^{-1}.

The correlation coefficients, $r = \sqrt{R^2}$ where R^2 is the coefficient of determination from linear regression of the EE to the total arsenic concentration, were compared with tabulated critical values for r. In case of Canada goldenrod (Fig. 4.5B), the calculated value of r plus the negative slope of the linear regression line suggested a significant inverse relationship between EE and total arsenic concentrations in the plant; that is, the more arsenic that was in the plant (Canada goldenrod) the less could be extracted. For the field horsetails (Fig. 4.5A), the purple loosestrife (Fig. 4.5E), and all (field and smooth) horsetails together (Fig. 4.5F) EE appeared not to be affected by the increase in the total arsenic content of the plants. The strong trends of increased and decreased EE with total arsenic were observed for panicle aster and sensitive fern, respectively (Fig. 4.5C and Fig. 4.5D). These findings showed that the EE of arsenic from plants may be dependent on the kind of plant species being extracted. The reasons for the different trends in the EE with the total arsenic content of plants may be attributed to the differences in the physiology of plant species and their different assimilation of the absorbed arsenic (also discussed in Section 4.8.1). The direct demonstration of the dependence or independence of extraction efficiency on plant matrices, to the best of our knowledge, has not been previously reported.

As will be discussed later in Chapter 5, the variation of arsenic species in plants with time of growth in the season would also demonstrate the differences in the way plants assimilate the absorbed arsenic in their systems. In this study, it was also observed that the way plants distribute the inorganic arsenic species in their systems was independent of the total absorbed arsenic (see Section 4.8.1).

The plant cell vacuoles play important roles in sequestering the essential as well as the waste and toxic metabolites in the plant system. It is possible that not all plants assimilate arsenic in the cell vacuoles at the same rate and same way. Arsenic may be complexed and/or sequestered in different forms and manners. Also, all plant tissues are not built the same way; some are very labile to reagents and/or heat while others are not. For example, apples and carrots are labile as opposed to shrubs and grasses that are not. Since EE is dependent on the extractant solvent's accessibility to, and ability for breaking up, the complex forms of the analyte, these factors have significant influence on the efficiency of the solvent for arsenic extraction. Therefore, for the extraction of arsenic from terrestrial plants by a specific method, it appears that extraction of each plant species has to be considered individually.

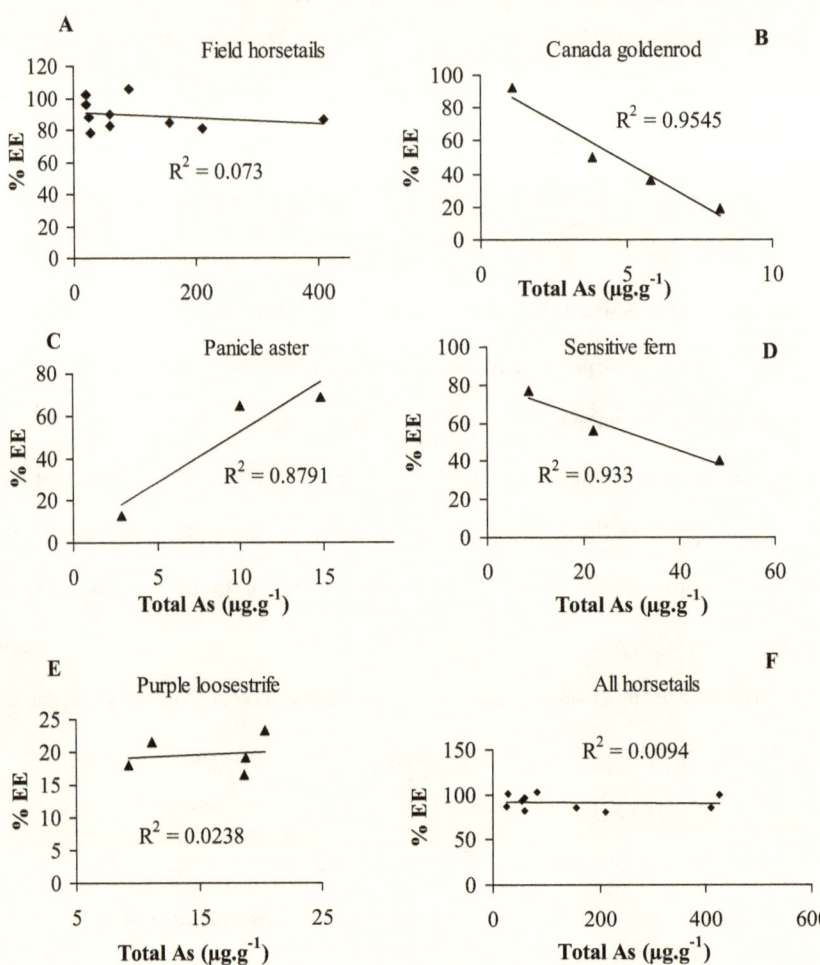

Figure 4.5. Extraction efficiency (EE) as function of total arsenic and plant matrix.

4.10. ARSENIC IN THE EQUISETUM (HORSETAIL) SPECIES

As was discussed earlier in Section 1.7.3, the *Equisetums* contained the highest concentrations of arsenic of all plants tested in this study (Section 4.5) and similar arsenic levels have been reported in other studies too.[120,121] The greatest amount of arsenic extracted was also from these plants and the %EE of these plants (n = 16) ranged from 78% to 108% with a median at about 95% (Tables 4.7, 4.8 and 4.9). The horsetails are members of the fern family,[148] known to contain a high amount of silica,

Arsenic in Plants: Extraction and Speciation

and are cultivated for pure silica.[118,156] They are used as food and for medicinal purposes.[114,117,157] Amounts of silica as much as 7.4 % - 9.8 % total, 2.4% water soluble and minimum 0.5% in true solution were reported.[116,158] In another study, it has been found that sonication yielded more colloidal silica in a short time from the plants.[118] Since the horsetails are used for food, medicinal and technical purposes and they are found to absorb high amounts of arsenic, an elaborate study of the plant is needed because of the risks that may arise from their use. On the other hand, the evaluation of their potential for use in the phytoremediation of the arsenic impacted areas is also important.

An analysis of the constituents of field horsetail with high protein and calcium content has been reported.[158] Arsenite binds to the thiolate groups of the proteins and enzymes of plant and animal tissues and forms complexes with the phytochelatins and glutathione.[129,153] Chemical interactions between arsenic, calcium and other essential elements in biological and aqueous systems are known.[159] Except for plants valued as foods, common plants are not known to contain significant amounts of amino acids or proteins.[160] The reasons for the *Equisetum*'s higher uptake of arsenic is not known; however, it may be related to its constitution of 18.6% amino acids (most likely proteins/enzymes) in the young bud of the plant and 400 mg Ca per 100 g in the adult plant.[158] On the other hand, high EE of arsenic (80-100%) can be attributed again to its constitution comprising amorphous silica that exists in loosely bound globules in the plant matrix,[161] and disperses in solution aided by sonication[118].

In this study, *Equisetum* species, particularly the field horsetails, showed significant absorption of arsenic. Amounts of arsenic in this plant collected from different time, soil and growth conditions varied from 19-530 µg.g^{-1}. Most of the absorbed arsenic could be extracted by the sequential method and 0.1 M-HCl single solvent extraction. A small but significant amount (4-8%) of MA, compared to the amounts of the other organoarsenic species found in the Deloro plants, was detected in this plant. It may be said, therefore, that a thorough study of the plant under controlled conditions may reveal important evidence regarding its arsenic absorption and assimilation processes and its potential for phytoremediation may be determined.

CHAPTER 5: EXTRACTION OF ARSENIC - FURTHER ASPECTS

5.1. INTRODUCTION

The development and application of the sequential arsenic extraction method has been described in the earlier chapters. The traditional extraction with water-methanol solvent has been augmented by the sequential use of dilute hydrochloric acid extraction aided by sonication. The results of sequential extraction of plants from Yellowknife, NWT and Deloro, ON in Canada have been reported in Chapter 3 and Chapter 4, respectively. A twofold improvement was demonstrated by the sequential extraction over the traditional water-methanol extraction in the overall extraction efficiency of arsenic from the terrestrial plants.

In this chapter, the results of further study on extraction efficiency of arsenic from plants by methanol, 1:1 water-methanol and 0.1 M-HCl sequential, and the single solvent 0.1 M-HCl are reported and discussed. The results of x-ray absorption near edge structure (XANES) experiments are presented. The results of HPLC-ICPMS confirming the presence of the organoarsenic species are also presented. The determination of multi-elements along with arsenic in the plants carried out by ICP-AES is also reported here.

5.2. THE WATER-METHANOL, SEQUENTIAL AND SINGLE SOLVENT ARSENIC EXTRACTION

The EE of sequential extraction by 1:1 water-methanol and 0.1 M-HCl has been reported in Chapter 3 and Chapter 4 for Yellowknife and Deloro plants, respectively, in Table 3.26 and Tables 4.7-4.10. The extractions of arsenic using solely 0.1 M-HCl from a selection of plants from Yellowknife and Deloro were carried out for comparison. The results from single solvent HCl extraction along with the EE from sequential process are reported in Table 5.1. The amounts of organoarsenic species found in the plants from 1:1 water-methanol extractions were added to the respective arsenic concentrations of plants determined from 0.1 M-HCl

Table 5.1. Extraction efficiencies of 1:1 water-MeOH (1:1), sequential and 0.1 M-HCl single solvent (SS) extractions. %EE of SS extractions after adding the organoarsenic species are also reported in the table for comparison.

Plant ID	Total As in plant (μg.g-1)	% EE, (1:1)	% EE, Sequential	% EE, (SS)	% EE, SS with organic species
YK-1*	95.2	9	30	28	28
YK-2	4.6	33	42	35	35
YK-3	18.5	10	36	39	39
YK-4*	8.0	19	38	50	50
YK-5	38.5	15	69	53	53
YK-6	47.5	52	105	101	101
YK-7	12.9	6	28	25	25
YK-8	4.4	27	28	41	41
YK-9	4.0	63	64	92	92
YK-10	8.1	46	46	60	60
YK-11*	75.8	4	31	29	30
YK-12	3.8	29	64	60	60
YK-13	3.0	48	105	79	79
DL-60*	24.9	74	88	88	88
DL-119*	5.8	29	36	74	78
DL-141*	9.3	6	18	76	77
DL-207*	157	32	85	87	90
DL-209*	20.3	3.8	23	57	58
DL-221*	26.3	37	78	95	103

*Plants contained organoarsenic species determined from sequential extraction.

The addition of the organoarsenic component of plants to the inorganic arsenic component of corresponding plants did not make a significant change in the overall EE due to the comparatively small contribution by the organic component. It should be noted, however, 0.1 M-HCl single extractions were as efficient as those of the sequential method for many plants and even better for a few plants. If the minor organoarsenic species were to be ignored a quicker extraction using only 0.1 M-HCl can be used in a one-step extraction.

The graphical representations of the EE of the sequential method and 0.1 M-HCl single solvent extraction are given in Figure 5.1 and Figure 5.2 for the Yellowknife and Deloro plants, respectively. As can be seen from the bar graphs the EE of the sequential method and 0.1 M-HCl single solvent extraction was similar for many plants, particularly those from Yellowknife (Fig. 5.1).

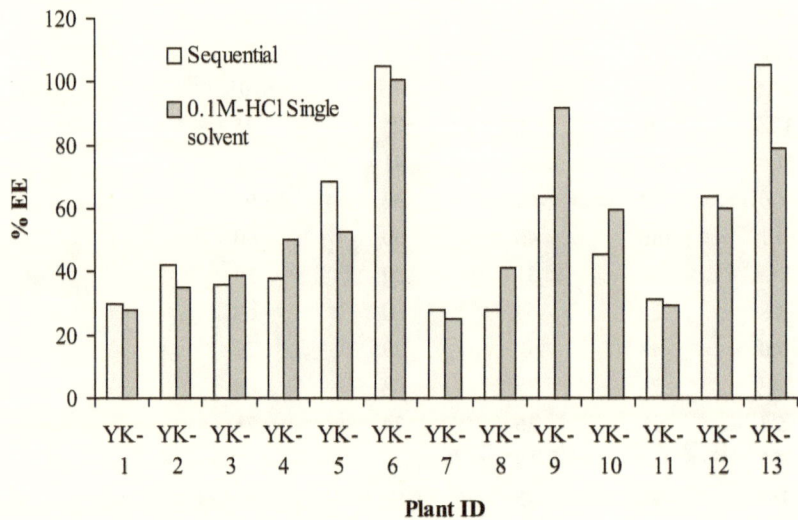

Figure 5.1. Comparison of EE by 1:1 water-methanol-0.1 M-HCl sequential and 0.1 M-HCl single solvent extractions of Yellowknife plants.

It was shown earlier that EE could be dependent on a number of factors, namely, the total arsenic in plants, plant matrices and the ratio of arsenite to arsenate in plants, and, as a consequence, a common trend of increased or decreased EE for the plants from 0.1 M-HCl single solvent extraction, and for that matter extraction with any solvent or method, was hard to achieve. For a few plants, DL-119, DL-141 and DL-209 in Figure 5.2, however, approximately 200-400% improved EE by 0.1 M-HCl was observed. In Section 5.4, the predominance of As(III) (>95% by XANES) in the purple loosestrifes (DL-141 and DL-209) is described. The Canada goldenrod (DL-119) appeared to act similarly. The extraction behavior of Canada goldenrod and purple loosestrife is shown in Figure 5.3 and appeared to be the same. This may be because both plants contain large amounts of As(III), however, the As(III) content in Canada goldenrod has not been determined by a method like XANES.

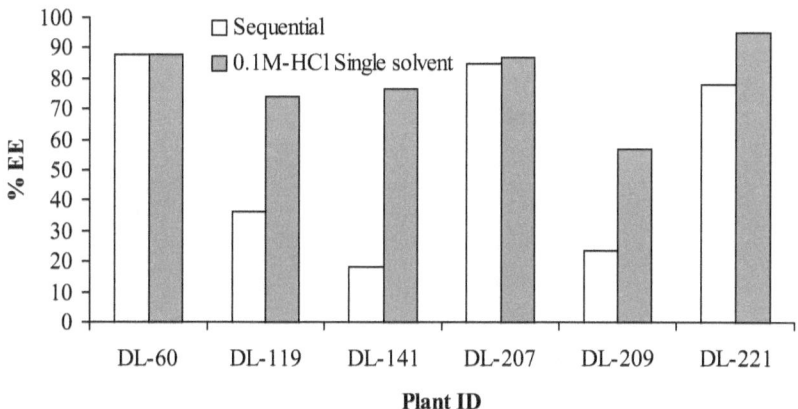

Figure 5.2. Comparison of EE by 1:1 water-methanol-0.1 M-HCl sequential and 0.1 M-HCl single solvent extractions of Deloro plants.

5.2.1. 100% Methanol Extraction of Arsenic in Plants

While reported in Chapter 3, the suppressing effect of methanol on arsenic extraction was further demonstrated from the extraction of Deloro plants by 50% and 100% methanol shown in Figure 5.4. The EE of 100% methanol was very low compared to the EE of 50% methanol extraction. Given 100% methanol suppresses the extraction of inorganic arsenic, as was demonstrated, the determination of organic arsenic compounds, particularly DMA, would be facilitated. It was shown earlier that DMA was better extracted in this medium (Chapter 3, Table 3.17).

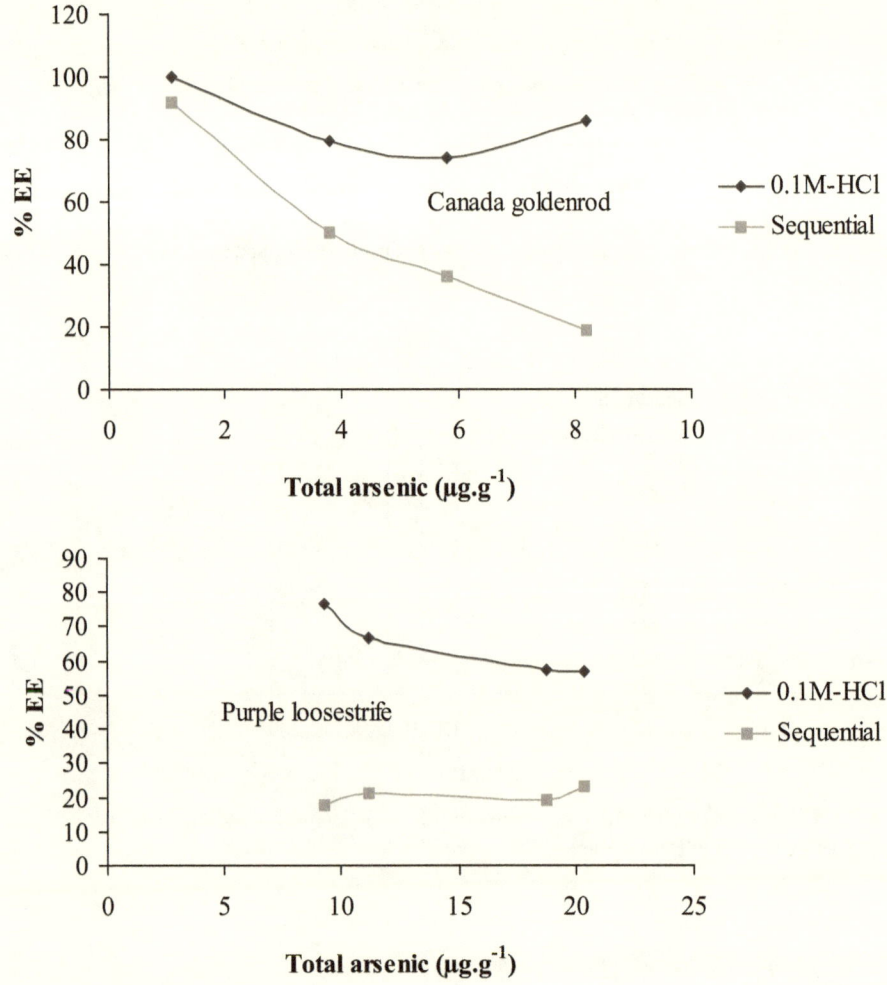

Figure 5.3. Extraction efficiencies of Canada goldenrod and purple loosestrife with respect to 0.1 M-HCl and sequential extractions.

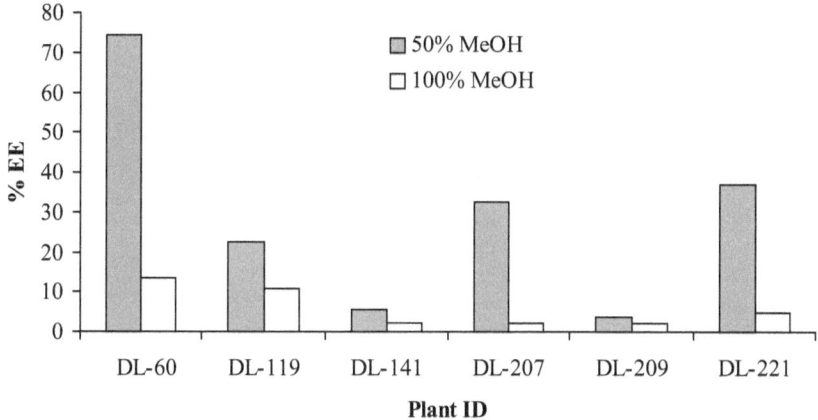

Figure 5.4. Extraction efficiencies of 50% methanol (1:1 v/v water-methanol) and 100% methanol.

From the results of extraction of arsenic from various plants and by different solvents, different schemes for arsenic extraction were devised. The three extraction schemes of arsenic comprised of a 100% methanol and 0.1 M-HCl sequence (Scheme I), 50% methanol (1:1 v/v water-methanol) and 0.1 M-HCl sequence (Scheme II) and 0.1 M-HCl single solvent extraction (Scheme III) are compared in Table 5.2. In Scheme I, methanol was completely removed by heating the residues from 100% methanol extraction prior to sequential extraction by 0.1 M-HCl. Removal of 100% methanol was necessary in order to minimize interference of methanol in the extraction of inorganic arsenic by the HCl medium. Scheme I is expected to aid DMA extraction in plants.

Scheme II is a sequential extraction by 0.1 M-HCl following 1:1 water-methanol extraction, as usual, without drying the residues from the water-methanol extraction. The extraction schemes have individual targets to achieve. Scheme II was developed in this study to analyze both organic and inorganic arsenic with better EE than the traditional water-methanol method. A detailed description of this scheme has been given in the preceding chapters. Extraction Scheme III, which is extraction with 0.1 M-HCl only, was intended for the faster assessment of the more toxic and prevalent inorganic arsenic species.

Table 5.2. Three extraction schemes of arsenic from plants. Extraction results of 100% methanol and 0.1M-HCl, 50% methanol and 0.1 M-HCl, and 0.1 M-HCl as single solvent (SS) extractants are compared. Arsenic concentrations are in $\mu g \cdot g^{-1}$.

Plant ID	Total As	Extraction scheme I			Extraction scheme II			Extraction scheme III	
		100% Methanol [A]	0.1 M-HCl [B]*	%EE, [A + B]	50% Methanol [C]	0.1M-HCl [D]	%EE, [C + D]	0.1 M-HCl (SS)	%EE, (SS)
DL-60	24.9	3.37	17.2	83	18.5	3.3	88	21.8	88
DL-119	5.8	0.63	2.7	57	1.3	0.8	36	4.3	74
DL-141	9.3	0.21	5.4	60	0.5	1.2	18	7.1	76
DL-207	157	3.45	130	85	51.2	81.9	85	136	87
DL-209	20.3	0.44	9.7	50	0.8	4.0	23	11.5	57
DL-221	26.3	1.31	21.1	85	9.7	10.8	78	25.0	95

*Methanol was completely removed before HCl extraction by heating residues of 100% methanol extraction in oven at about 60°C.

5.3. THE ORGANOARSENIC COMPOUNDS IN PLANTS: COMPARISON OF HPLC-HGAAS AND HPLC-ICPMS RESULTS

A number of plant extracts from sequential extraction which showed the presence of organoarsenic species and were quantitatively determined by HPLC-HGAAS were further studied by HPLC-ICPMS. Both anion and cation exchange HPLC columns were used in connection with the ICPMS. The experimental conditions have been described in Chapter 2. The arsenic standards As(III), DMA, MA and As(V) were used in an anion exchange column and AB, TMAO and Tetra in a cation exchange column. The species, DMA and MA, which were determined by HPLC-HGAAS, were confirmed by the HPLC-ICPMS experiments. The results of quantitative analysis of the arsenic species in plant extracts by HPLC-HGAAS and HPLC-ICPMS are reported in Table 5.3 for comparison. The results of the arsenic species determined by two methods agreed very well. The overall precision of the two analytical techniques was comparable. The sensitivity of HPLC-ICPMS is at least an order of magnitude higher than the sensitivity of HPLC-HGAAS[162] as is reflected in the determination of DMA in 50% methanol extract of plant DL-200 and in the determination of As(V) in 100% methanol extracts of plants DL-209 and DL-220. While the detection of these arsenic species in the extracts was possible by HPLC-ICPMS, the concentrations of the species were below the LOD of HPLC-HGAAS.

Table 5.3. Comparison of HPLC-HGAAS (HGAA) and HPLC-ICPMS (ICP) results of arsenic species in 100% methanol and 50% methanol extracts of Deloro plants. <LOD indicates concentration of species was close but not above detection limit. All concentrations are in $\mu g.g^{-1}$.

Extraction solvent	Sample ID	As(III) HGAA	As(III) ICP	DMA HGAA	DMA ICP	MA HGAA	MA ICP	As(V) HGAA	As(V) ICP
100% Methanol	DL-105	nd	nd	0.51	0.73	nd	nd	1.06	1.04
	DL-119	nd	nd	0.26	0.34	nd	nd	0.29	0.30
	DL-119 Dupl.	nd	nd	0.37	0.35	nd	nd	0.34	0.29
	DL-129	nd	nd	0.18	0.27	nd	nd	nd	nd
	DL-207	0.096	0.107	nd	0.011	0.86	0.92	2.6	3.1
	DL-207 Dupl.	0.049	0.097	nd	0.011	0.85	0.84	2.5	2.7
	DL-209	nd	0.041	0.29	0.34	nd	nd	<LOD	0.07
	DL-220	nd	0.033	0.23	0.31	nd	nd	<LOD	0.04
50% Methanol	DL-200	nd	nd	<LOD	0.04	nd	nd	2.51	2.57
	DL-201	0.18	4.12	nd	nd	nd	nd	7.64	5.10
	DL-207	6.51	0.94	nd	nd	4.56	4.69	40.1	43.4
	DL-207 Dupl.	7.15	3.10	nd	nd	4.61	4.18	36.3	36.9
	DL-209	nd	nd	0.28	0.15	nd	nd	0.50	0.65
	DL-119	nd	nd	0.23	0.42	nd	nd	1.07	0.87

Representative HPLC-ICPMS chromatograms are presented in Figure 5.5. Any species other than those detected by HPLC-HGAAS were not detected in any significant amount by the HPLC-ICPMS. There were no cationic species (AB, TMAO or Tetra) detected in any of the samples tested by the cationic HPLC-ICPMS experiments. These observations show that the use of an anion exchange column connected with HGAAS would be sufficient in case of routine commercial analysis of arsenic in plant samples particularly those grown on arsenic contaminated soils.

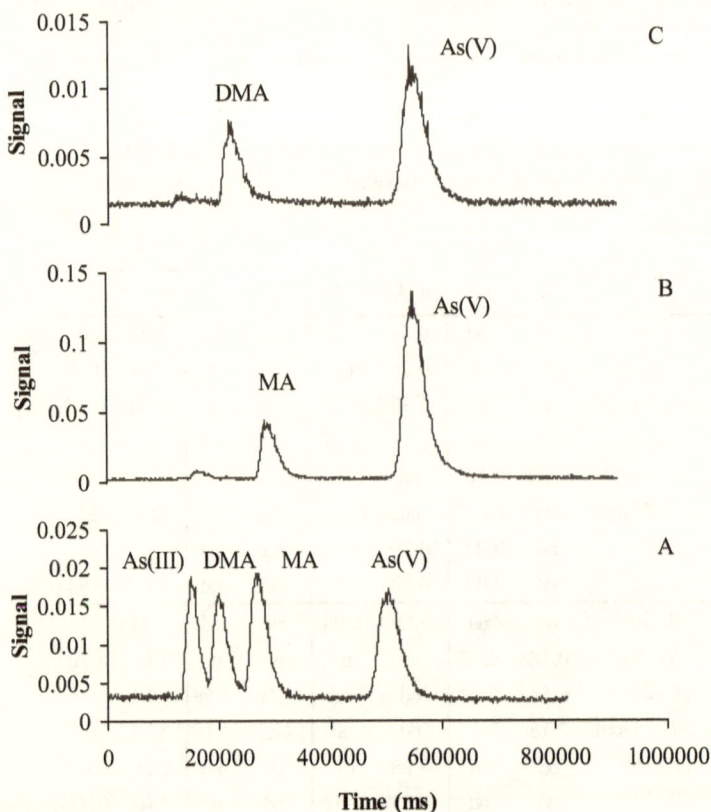

Figure 5.5. HPLC-ICPMS anion exchange chromatograms of: (A) standard (20 μg.L^{-1}) solutions of arsenic compounds, (B) Deloro plant DL-207 and (C) DL-119. ICPMS confirmed MA in DL-207 and DMA in DL-119 determined by HPLC-HGAAS. Other organoarsenic species were not detected in these plants by either method.

5.4. XANES ANALYSIS OF DELORO PLANTS

A description of the x-ray absorption near edge structure (XANES) is provided in Chapter 2. Nineteen plant samples were sent to Argonne Synchrotron Laboratory in Illinois, USA for XANES analysis. The samples included fresh frozen plants (fresh plants frozen after sampling without drying), dry ground samples, and dried residues from the sequential extraction of plants. XANES LC-fit of the X-ray absorption spectra of plant samples with

As(III), As(V) and As(III)-Glu standard spectra were carried out using WinXAS program.[61] LC-fit is carried out by fitting the experimental XANES data to linear combination (LC) of the data from known reference compounds. The fractions of As(III) and As(V) in the plants were first estimated. Next, the fraction of As(III) bound to S (sulfur or thiol group) in plant was determined along with the other As(III) and As(V) fractions. A few representative XANES spectra of standard As(III) and As(V) and plant samples are shown in Figure 5.6 and Figure 5.7, respectively.

The XANES LC-fit results of the arsenic species in the plants are reported in Table 5.4. The total percentages for As(III) and As(V) in nineteen plants ranged from 97% to 126% with a median at 102%. For As(III), As(V) and As(III)-Glu the range was 98% to 206% with a median at 103%. The arsenic species are distinguished by the difference in the white lines, i.e., the peaks of the X-ray absorption edge. The absorption edge energy between As(III) and As(V) is 3.6 eV which is larger than 1.7 eV between As(III)-(Glu) and As(III).[61,163]

Thus the speciation of As(III) and As(V) is simpler than the As(III), As(III)-Glu and As(V) species together. Since the white line energies of the As(III), As(III)-Glu and As(V) species are closely separated, a small difference in the X-ray absorption cross section of white line peaks for these species may alter the ratios of the species significantly. These were evident in the speciation of As(III), As(V) and As(III)-Glu together for the plant samples of this study. It appeared that the fraction of As(III) which lies in between X-ray absorption edges of As(III)-Glu and As(V), showed too high values for some samples (for example, purple loosestrife (220b and 220c in Table 5.4). However, it is important to realize that due to many factors (e.g., X-ray self-absorption, large differences between standard and analyte concentrations, and selection of XANES data values) involved in the entire XANES process the fractions obtained from XANES fit were qualitative and need to be interpreted as such.[61]

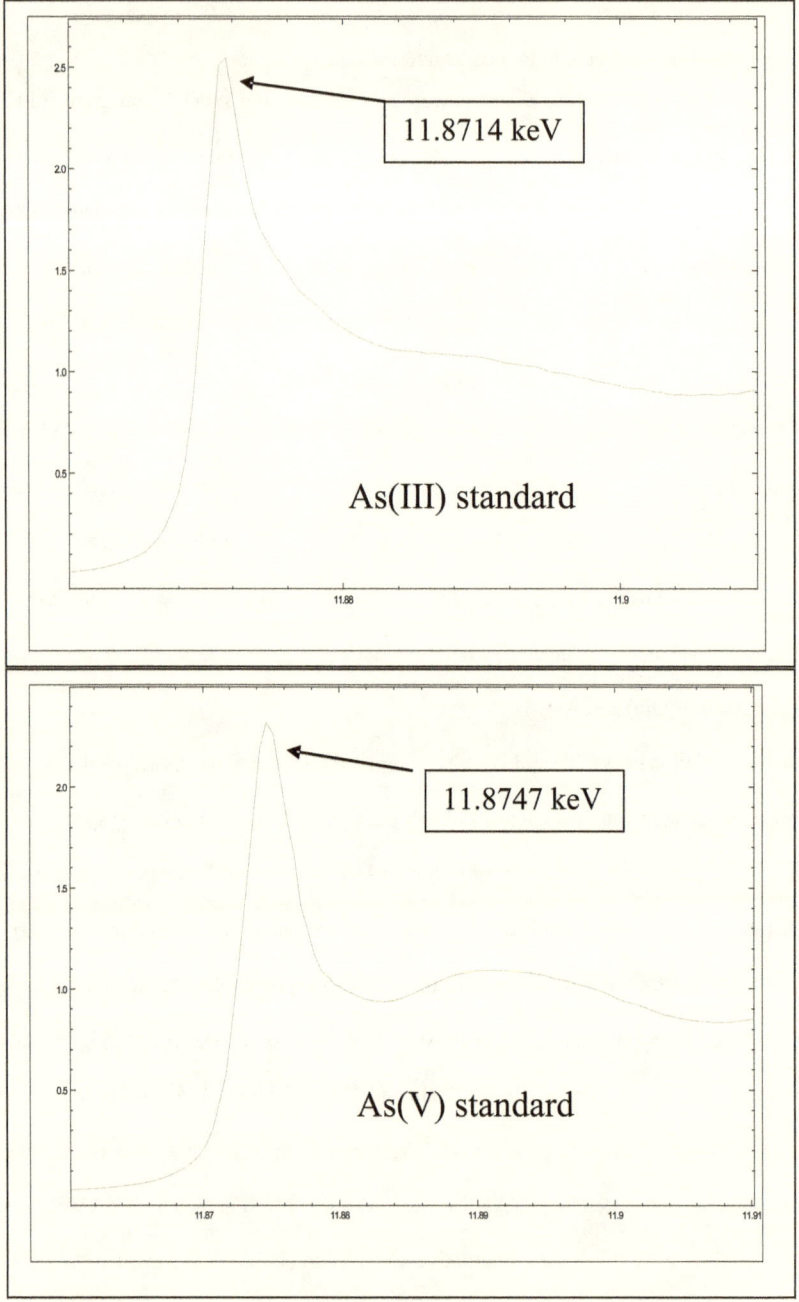

Figure 5.6. XANES spectra (Photon energy (keV) vs. Normalized absorption (a. u.)) of standard As(III) and As(V) species.

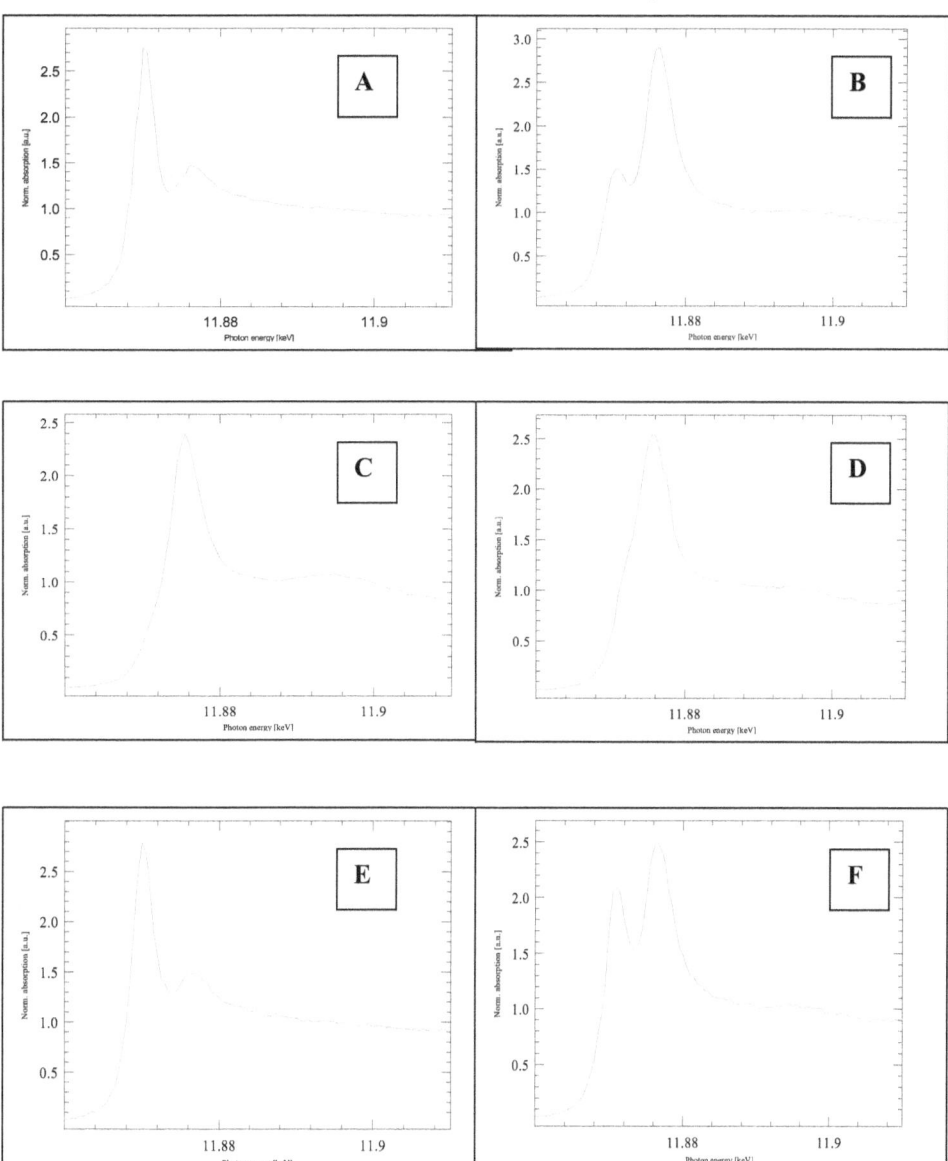

Figure 5.7. XANES spectra (Photon energy (keV) vs. Normalized abs. (a. u.)) of dry ground plant samples and corresponding residues from sequential extractions are presented. A: Dry ground sensitive fern, B: Residue of sensitive fern; C: Dry ground field horsetail, D: Residue of field horsetail; and E: Dry ground purple loosestrife, F: Residue of purple loosestrife.

Kalam Mir

The sensitive fern and purple loosestrife samples contained As(III) that was more than 90% of the total As in all samples in the XANES analysis as reported in Table 5.4. Arsenite fractions in the three field horsetails (DL-15, DL-101 and DL-207a) varied from 20 to 96% with the median at 63%. The As(III) fraction in the smooth horsetail (DL-50) was 97%. In the Yellowknife plants, slender wheat (YK-1) and blue joint (YK-11) grasses, amounts of As(III) and As(V) were approximately 40% and 60%, respectively. Variable amounts of sulfur bound arsenite, represented by the As(III)-Glu fractions, ranging from 0-62% was also observed in all plants. In the reported studies, the stability of arsenic-glutathione complexes in solution has been observed[164] and primary response of plants to this metalloid toxicant was the formation of arsenic-glutathione complexes or phytochelatins has been demonstrated.[85, 129,131]

The advantages of XANES are due to its capability of analyzing both solid and liquid samples directly in their original states without performing the often-lengthy extractions and conventional speciation processes which may alter the speciation of the element in the original sample. This gives the opportunity for comparison of the species in the original sample and extracted species. However, XANES sensitivity for elements is low compared to the other analytical methods (ppm vs. ppb), such as the HPLC-HGAAS and HPLC-ICPMS.

5.4.1. Arsenic Species in Fresh and Dried Ground Plants

The percentages of arsenite and arsenate in the fresh frozen and dry ground parts of three plants are shown in Figure 5.8. The three plants tested are different species; sensitive fern (DL-201), field horsetail (DL-207), and purple loosestrife (DL-220). For all three plants, the arsenite and arsenate fractions essentially remained unchanged during washing, drying and grinding processes. This showed that the As(III) and As(V) pair in these plants were stable. The stability of the As(V) and As(III) has been discussed in Chapter 1.4.1. The XANES results for the As(V) and As(III) fractions are from two different environments; one is the aqueous environments of cellular tissues of the undried plant matrix and the other is the dried and ground plant matrix. This indicates the good stability of the redox pair.

Table 5.4. XANES analysis of Deloro and Yellowknife plants. Results from XANES fit with As(III) and As(V), and As(III), As(V) and As(III)-Glu together are reported. ND indicates values insignificant or negative.

Plant ID	Plant name	Sample state	Total As ($\mu g \cdot g^{-1}$)	XANES Fit (% species)				
				As(III)	As(V)	As(III)	As(V)	As(III)-Glu
DL-15	Field horsetail	Dry Ground	530	96	11	97	9	1
DL-50	Smooth horsetail	Dry Ground	27.6	97	7	74	6	22
DL-101	Field horsetail	Dry Ground	410	63	38	47	38	16
DL-104	Sensitive fern	Dry Ground	22	93	8	43	10	50
DL-128	Purple loosestrife	Dry Ground	19	97	5	70	3	29
DL-201a	Sensitive fern	Dry Ground	48	98	2	36	5	62
DL-201b		Residue	29	59	60	68	61	ND
DL-201c		Fresh Frozen	Unknown	92	9	58	4	40
DL-207a	Field horsetail	Dry Ground	157	20	86	23	88	ND
DL-207b		Residue-1	23	40	70	130	68	ND
DL-207c		Residue-2	23	40	71	138	68	ND
DL-207d		Fresh Frozen	Unknown	21	85	20	84	ND
DL-220a	Purple loosestrife	Dry Ground	19	100	1	56	ND	52
DL-220b		Residue	15	86	28	177	19	ND
DL-220c		Fresh Frozen	Unknown	126	ND	193	ND	ND
YK-1a	Slender wheat grass	Dry Ground	95	42	55	15	45	39
YK-1b		Residue	67	55	44	24	37	40
YK-11a	Blue joint grass	Dry Ground	77	38	64	35	60	7
YK-11b		Residue	54	56	46	44	36	21

Arsenic in Plants: Extraction and Speciation

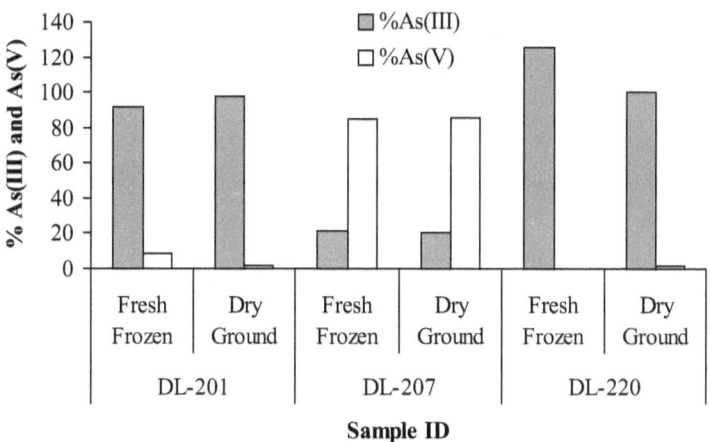

Figure 5.8. As(III) and As(V) in fresh frozen plant and dry ground sample by XANES analysis. DL-201: sensitive fern (*Onoclea sensibilis*); DL-207: field horsetail (*Equisetum arvense*) and DL-220: purple loosestrife (*Lythrum salicaria*).

5.4.2. Arsenic Species in Dry Ground Plant samples and Corresponding Residues of Extraction: Implication for Extraction Efficiency (EE)

A number of residues from sequential extraction as listed in Table 5.4 along with the original dry ground samples of field horsetail (207a and 207b/207c), sensitive fern (201a and 201b), purple loosestrife (220a and 220b), slender wheat grass (YK-1a and YK-1b) and blue joint grass (YK-11a and YK-11b) underwent XANES analysis. The results of As(III) and As(V) from the XANES analysis are reported in graphical forms in Figure 5.9. The plants that were hard to extract for arsenic, and the field horsetail, for comparison purposes, were chosen for the XANES analysis. The %EE of the plants were 40, 19, 29, 35, and 85 for sensitive fern, purple loosestrife, slender wheat grass, blue joint grass, and field horsetail, respectively.

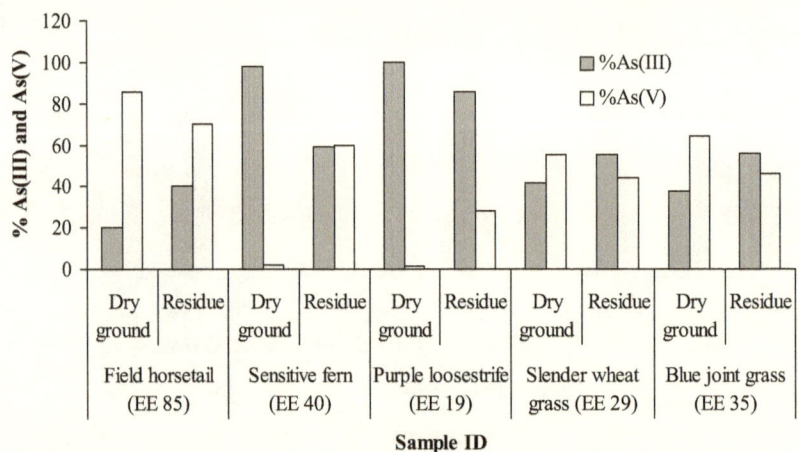

Figure 5.9. Comparison of fractions of As(III) and As(V) in original dry ground plant samples and residues from sequential extractions. Extraction efficiency of each plant species is given in parentheses.

Except for the sensitive fern, the proportions of As(III) and As(V) fractions were similar in the dry ground and corresponding residues of the plants. The residues of sensitive fern and purple loosestrife showed increase in the As(V) fraction compared to that in the original dry ground samples. This increase may indicate that the trapped As(III) in plant matrix is somehow being converted to As(V) in the extraction and speciation processes.

5.4.3. Variation of Arsenic Species in Season

A number of plant samples were evaluated in terms of their sampling time at Deloro in May, July and September. The results from XANES analysis for As(III), As(V) and As(III)-Glu fractions in these plants were compared. The May, July and September fractions of the arsenic species in the three plants are shown in bar graphs in Figure 5.10.

Arsenic in Plants: Extraction and Speciation

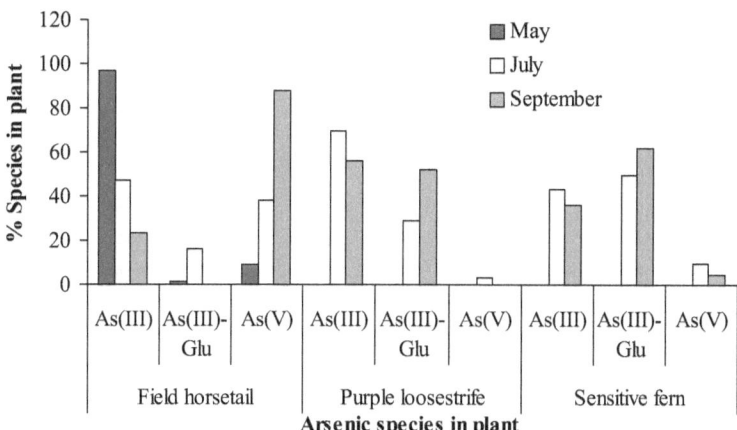

Figure 5.10. Variations in the species of arsenic in a number of plant species during a growing season.

For the field horsetail, the As(III) fraction was the highest in the May plant and lowest in the September plant; and a reverse trend for As(V) was observed. A small fraction of As(III)-Glu was present in the July plant of this species but little or none of this was present in September. This horsetail showed a gradual decrease of As(III) and increase of As(V) over time in the season. Horsetails generally have high EE as well as the ability to uptake higher amounts of arsenic than most other species. This data shows a very different pattern of arsenic species in horsetail over the seasons.

The purple loosestrifes and sensitive ferns were sampled in July and September only since May was early in the season for the plants to be sampled for this study. In both plants, As(III) and As(III)-Glu were the predominant species throughout the season. A trend in the increase of the sulfur bound arsenic, that is the As(III)-Glu fraction, was observed at the end of the season in September plants while As(V) was always a very minor species in these plants. It should be noted here that these plants, the purple loosestrife and sensitive fern, also showed low average (mean ± average deviation) extraction efficiencies of $(18 \pm 2)\%$ and $(48 \pm 8)\%$, respectively, and showed some As(III) to As(V) conversion in the residual plant matrix (Fig. 5.8)

5.4.4. XANES and HPLC-HGAAS Speciation of Arsenic

The As(III) and As(V) fractions in the dry ground samples of these plants determined by XANES were compared with those in their sequential extracts determined by HPLC-HGAAS. The percentage fractions of both species from XANES and HPLC-HGAAS analyses along with the extraction efficiencies (EE) of the plants by sequential method are presented in Table 5.5. Because of the variable extraction efficiencies of the plants ranging from 16% to 100%, a straightforward comparison of the results from XANES and HPLC-HGAAS was difficult. Besides, the sensitivities of the two analytical methods are widely different. The percent fractions of the two arsenic species in dry ground plant samples of DL-101, DL-128, DL-207 and DL-220 determined by XANES were comparable with those in their extracts determined by HPLC-HGAAS.

Table 5.5. The results of As(III) and As(V) from XANES (fractions in solid ground sample) and HPLC-HGAAS (fractions in liquid extracts) analysis are reported.

Plant ID	Plant name	Total As ($\mu g \cdot g^{-1}$)	% from XANES		% from HPLC-HGAAS		%EE, Sequential
			As (III)	As (V)	As(III)	As(V)	
DL-50	Smooth horsetail	28	97	7	46	54	100
DL-101	Field horsetail	410	63	38	48	52	86
DL-104	Sensitive fern	22	93	8	41	59	56
DL-128	Purple loosestrife	19	97	5	97	3	16
DL-201	Sensitive fern	48	98	2	50	50	40
DL-207	Field horsetail	157	20	86	10	90	85
DL-220	Purple loosestrife	19	100	1	84	16	19
YK-1	Slender wheat grass	95	42	55	12	88	28
YK-11	Blue joint grass	77	38	64	4	96	24

Arsenic in Plants: Extraction and Speciation

For the other plants, DL-50, DL-104, DL-201, YK-1 and YK-11, an inspection of the As(V) with As(III) fractions of XANES and HPLC-HGAAS showed that the As(V) fractions in the HPLC-HGAAS were higher. This suggested that some of the As(III) in the original sample was being converted to As(V) in the extraction process that is clearly evident in case of DL-50. The sources of the increased As(V) whether As(III) or As(III)-Glu or both were not clear since these plants showed significant fractions of both species in their matrices. However, since earlier speciation experiments using spikes showed stability of both species, As(III) and As(V), in the extraction processes, a likely alternative would be the As(III)-Glu species. Further investigations involving a large number of plant samples and the arsenic species including As(III)-Glu will be necessary to understand the speciation process.

5.5. ARSENIC AND OTHER ELEMENTS IN PLANTS GROWN ON ARSENIC IMPACTED SOILS

Multi-element determination in plants was carried out by using the multi-element analytical capability of the ICPAES. Thirty elements including arsenic were determined in 22 plant samples. The elements were Ag, Al, As, B, Ba, Be, Ca, Cd, Co, Cr, Cu, Fe, K, Mg, Mn, Mo, Na, Ni, P, Pb, S, Sb, Se, Sn, Sr, Ti, Tl, U, V and Zn. Among the elements, Be, Sb, Se and Tl were the least abundant and not detected in most of the plants. Other than C, H, O and N, the major elements that have been determined in plants are Ca, K, Mg, P and S.[165] The latter elements which are termed as the macro-elements of plants were also found to be the most abundant in this study. The concentrations of the macro-elements along with other elements including arsenic are reported in Table 5.6.

In the plants, K existed at similar levels (median about 21,600 $\mu g.g^{-1}$ and range about 33,700 $\mu g.g^{-1}$) in most of the plants. For Mg and P, medians and ranges were approximately 2,000 $\mu g.g^{-1}$ and 2,700 $\mu g.g^{-1}$, and 1,700 $\mu g.g^{-1}$ and 4,400 $\mu g.g^{-1}$, respectively. Both Ca and S reached their highest levels of concentration, 25,000-28,000 $\mu g.g^{-1}$ and 11,000-17,000 $\mu g.g^{-1}$, respectively, at the highest arsenic levels (221-405 $\mu g.g^{-1}$) in plants. Their medians and ranges were approximately, 13,000 $\mu g.g^{-1}$ and 24,000 $\mu g.g^{-1}$, and 3,000 $\mu g.g^{-1}$ and 16,000 $\mu g.g^{-1}$, respectively.

The heavy metals including arsenic determined in the plant samples were separated into three groups according to the plant physiological demands such as the essential, not essential but considered as beneficial, and those non-essential but potentially toxic elements.[166]

Table 5.6. Results of arsenic, macro and minor elements in Deloro plants are reported. All concentrations are in µg.g^{-1}.

Plant ID	Plant name	As	Al	B	Ba	Ca	K	Mg	Na	P	S
DL-44	Bur oak	0.24	12	37	12	7470	11400	2010	53	4000	1080
DL-52	Poplar balsam	0.82	10	28	12	6070	17600	2140	50	3520	1290
DL-18	Canada goldenrod	2.4	27	45	20	10300	39100	2180	103	5310	2080
DL-46	Trout lily	2.8	16	25	7	3880	15300	1470	220	1270	1420
DL-47	Spruce	3.0	29	22	81	8220	9670	666	48	1340	736
DL-205	Panicle aster	3.3	16	63	19	15000	26300	2490	54	4420	5910
DL-49	Common juniper	3.9	56	22	11	10300	7220	830	57	838	569
DL-119	Canada goldenrod	6.2	14	65	10	15900	25600	3700	65	2990	1990
DL-133	Ox-eye daisy	7.7	11	47	15	11300	29100	1950	98	1810	2530
DL-129	Canada goldenrod	8.4	14	94	14	14100	27400	3760	60	2030	1180
DL-6	Sensitive fern	9.8	9	32	51	3500	21100	2230	100	4370	2300
DL-225	Panicle aster	10	15	66	5	11400	23200	2140	64	2490	4110
DL-200	Wild strawberry	10	20	38	227	14900	6370	4740	65	1660	1030
DL-103	Bracken fern	11	14	20	177	3800	11400	2970	45	1660	775
DL-143	Panicle aster	16	11	85	3	9760	30500	2510	67	1350	4530
DL-209	Purple loosestrife	21	23	28	26	21200	18800	4250	152	1220	8830
DL-135	Smooth horsetail	41	7	33	6	12200	30000	2240	300	1240	10100
DL-212	Smooth horsetail	43	11	38	6	16700	25700	2190	278	1330	8780
201 dup	Sensitive fern	52	28	33	26	17200	5560	6910	53	1340	8060
DL-201	Sensitive fern	52	31	35	20	17100	5460	7100	64	1380	8270
DL-56	Field horsetail	98	6	35	4	13700	32400	3670	105	4020	7950
DL-130	Field horsetail	223	12	40	34	28200	21600	4910	93	1320	16600
DL-101	Field horsetail	406	31	39	58	25300	26400	3510	138	2140	10800

Table 5.7. Arsenic and other essential, beneficial and nonessential elements in plants are listed. All concentrations are in $\mu g \cdot g^{-1}$.

Plant ID	Plant name	As	Cu	Fe	Mn	Mo	Zn	Co	Ni	V	Cd	Cr	Pb	Tl
DL-44	Bur oak	0.24	14	58	43	nd	34	0.74	2.1	0.15	0.06	0.36	12	nd
DL-52	Poplar balsam	0.82	12	40	32	nd	112	3.1	2.6	0.14	0.26	0.38	10	nd
DL-18	Canada goldenrod	2.4	17	85	34	1.7	26	9.6	12	0.23	0.02	0.87	7.0	nd
DL-46	Trout lily	2.8	9	48	11	0.30	17	0.59	0.4	0.11	0.16	0.52	30	nd
DL-47	Spruce	3.0	7	43	58	nd	102	0.20	1.0	0.20	0.14	0.29	26	nd
DL-205	Panicle aster	3.3	42	59	58	1.6	87	16	5.0	0.25	0.13	0.46	8.2	nd
DL-49	Common juniper	3.9	11	67	26	nd	15	0.09	1.3	0.33	0.17	0.44	23	nd
DL-119	Canada goldenrod	6.2	19	96	53	1.8	14	7.5	14	0.24	0.15	14	22	nd
DL-133	Ox-eye daisy	7.7	22	36	24	0.22	71	1.5	4.1	0.16	0.25	0.3	9.1	nd
DL-129	Canada goldenrod	8.4	19	56	40	0.08	31	2.5	2.9	0.22	0.03	0.4	14	nd
DL-6	Sensitive fern	9.8	20	57	43	nd	33	17	5.7	0.09	0.32	0.3	27	nd
DL-225	Panicle aster	10	33	50	36	1.5	86	0.43	1.4	0.18	0.08	0.3	18	nd
DL-200	Wild strawberry	10	15	47	206	nd	28	2.4	2.7	0.25	0.18	0.4	16	nd
DL-103	Bracken fern	11	8	62	146	nd	20	4.8	0.9	0.13	0.07	0.5	5.0	nd
DL-143	Panicle aster	16	36	50	133	3.4	53	1.5	3.6	0.14	0.20	0.3	19	nd
DL-209	Purple loosestrife	21	13	61	19	1.1	309	5.5	3.0	0.35	0.03	0.4	10	nd
DL-135	Smooth horsetail	41	8	38	41	nd	66	3.4	0.7	0.19	0.16	0.4	6.5	nd
DL-212	Smooth horsetail	43	5	35	57	nd	73	4.9	0.7	0.25	0.06	0.6	1.5	nd
201 dup	Sensitive fern	52	9	234	109	1.3	14	28	20	0.82	0.02	21	2.4	nd
DL-201	Sensitive fern	52	9	154	108	0.8	15	28	12	0.79	0.02	0.6	1.8	nd
DL-56	Field horsetail	98	12	53	26	nd	41	0.50	0.3	0.23	0.07	0.8	5.0	nd
DL-130	Field horsetail	223	15	45	30	0.19	25	3.5	1.1	0.40	0.04	1.8	0.84	nd
DL-101	Field horsetail	406	12	230	203	4.3	38	66	25	0.37	0.03	1.3	1.6	0.25

Arsenic in Plants: Extraction and Speciation

The concentrations of these elements are reported in Table 5.7. Due to the diversity in the plant species, metal concentrations in the plants varied widely. Arsenic concentration in the plants ranged between 0.2-406 µg.g^{-1} and Mo, V and Cd were the least concentrated metals ranging 0.1-4.3, 0.09-0.82 and 0.02-0.32 µg.g^{-1}, respectively.

In published studies, correlation between concentrations of arsenic and other heavy metals in arsenic contaminated soils has been reported.[167] However, such correlations may be influenced by different environmental factors such as area topology, climatic conditions, vegetation and land use. The presence of inter-element correlation in reference plants and study samples has also been reported.[168]

5.5.1. Correlation of Arsenic with Sulfur and Calcium in Plants

Since arsenic is known to affect the physiologies of both plants and animals, a study was conducted to determine how the overall concentrations of physiologically essential, beneficial and non-essential elements correlated with the arsenic concentrations in plants grown on arsenic contaminated soils. Inter-element correlation in plants was observed;[168] however, in the current work the correlation between arsenic and other elements was studied.

From the results of the current study, and among the elements reported in Table 5.6, the concentrations of Al, B, Ba, K, Mg, and Na showed very weak to weak correlation with the concentration of As in plants indicating that the levels of these elements in plants remained steady irrespective of the low or high arsenic concentrations. The range of correlation coefficient (r) from regression analysis was 0.024-0.288 which is smaller than the critical value, $r_{0.05,\ 20} = 0.423$. However, stronger correlations were observed for the plots of As, S and As, Ca pairs as shown in Figure 5.11. Second order polynomial fits gave correlation coefficients of 0.885 and 0.757 for the As, S and As, Ca pairs, respectively. The correlations were found to be stronger for both S and Ca at the lower level arsenic concentrations (0.2-52.1 µg.g^{-1}). These correlation coefficients are larger than the critical value and suggest a significant association between the concentrations of arsenic and the concentrations of Ca and S in the plants. This correlation was absent for the other macro-elements in the plants. To our best knowledge, this is the first reported evidence of a positive correlation between the absorbed arsenic with sulfur and calcium in the plants grown on arsenic contaminated soils.

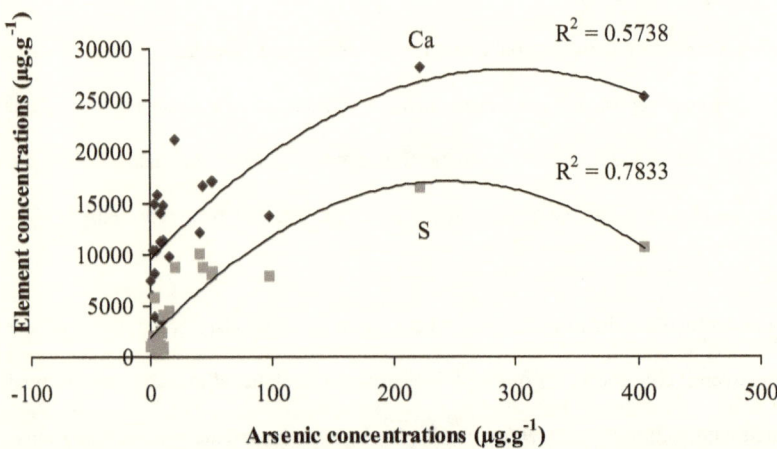

Figure 5.11. Plots of arsenic concentrations vs. calcium and sulfur concentrations in plants from Deloro.

A correlation between arsenic and sulfur in plants was expected. It has been shown that both As(III) and As(V) efficiently induced the biosynthesis of phytochelatins ([γ-glutamate-cysteine]$_n$-glycine) a substance synthesized from sulfur containing reduced glutathione (GSH) in plants, and that arsenic-phytochelatins (As-PC) were present in weak acid extracts.[129,169] As(III) is found to coordinate with sulfur (S) of the thiol groups (SH) of PC in various proportions.

The XANES analysis in present study showed presence of As(III)-Glu fractions in a number of plants as described above. In Figure 5.11, the slopes of the curves for S and Ca versus As attained their maxima at around 250-300 µg.g^{-1} arsenic concentrations in plants and then slid toward negative values. This may be indicative of the fact that plants could keep up with arsenic absorption to a moderate level by chemically binding it with the sulfur containing PC and calcium compounds within the plant matrix.

In Figure 5.11, R^2 values for a second order polynomial are shown. R^2 values for linear fits (0.4705 and 0.4502 for As-Ca and As-S, respectively) still show a relation with values above critical levels.

5.5.2. Correlation between Arsenic and other Heavy Metals in Plants

The concentrations of arsenic and other heavy metals presented in Table 5.7 were studied for evaluating the relationship between arsenic and the other metals in plants. Very weak to moderate but below critical level correlation was observed for most of the metals. The essential metals Fe ($r = 0.546$) and Mo ($r = 0.496$) showed overall correlation above critical level with positive slope of the linear regression. This correlation was found to be stronger for Fe ($r = 0.574$) at the lower level arsenic concentrations (0.2-52.1 $\mu g.g^{-1}$), Figure 5.12.

Manganese showed a moderate overall correlation ($r = 0.402$) with positive linear regression slope but was weak ($r = 0.271$) at the lower level arsenic concentrations.

Cobalt showed the highest correlation throughout arsenic levels ($r = 0.601$-0.732) with a positive slope of linear regression, Figure 5.13. Another beneficial element Ni showed overall correlation ($r = 0.523$) with positive slope above critical level, however, was weaker ($r = 0.384$) at the lower level of arsenic concentrations. Vanadium was weak in the overall correlation ($r = 0.286$) but showed much stronger correlation ($r = 0.732$) above the critical level ($r_{0.05,\ 17} = 0.456$) with positive slope at the lower level of arsenic concentrations.

Among the nonessential but potentially toxic elements Cd showed a moderate correlation ($r = 0.355$-0.376) throughout the arsenic concentration range with a negative slope of the linear regression line, Figure 5.14. The other toxic element Pb was very weakly correlated ($r = 0.073$) overall but at lower level arsenic the correlation was stronger with a positive slope ($r = 0.297$) still much below the critical level.

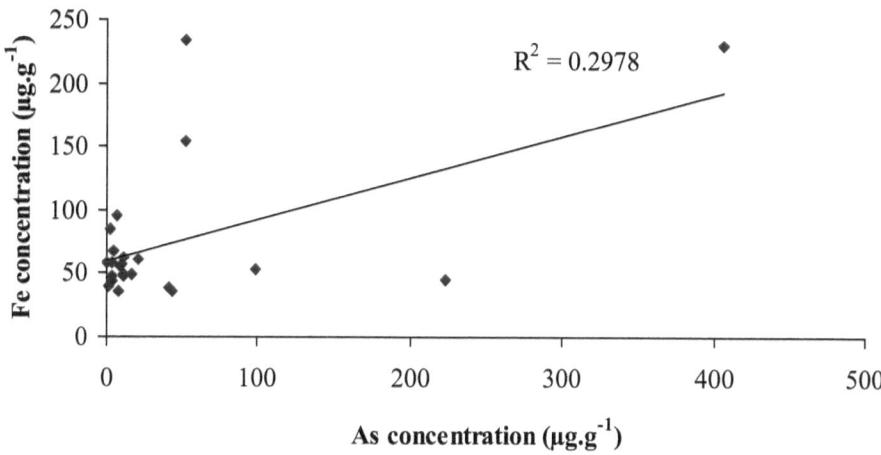

Figure 5.12. Cluster plot of arsenic and iron in plant samples.

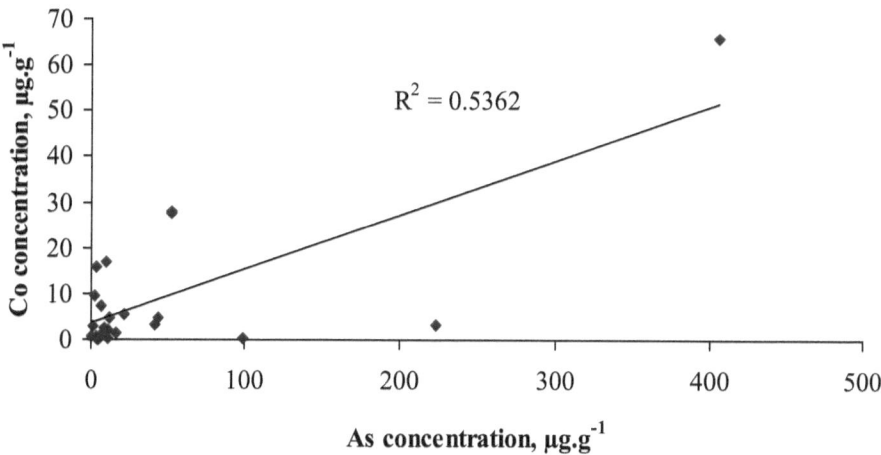

Table 5.13. Cluster plot of arsenic and cobalt in plant samples.

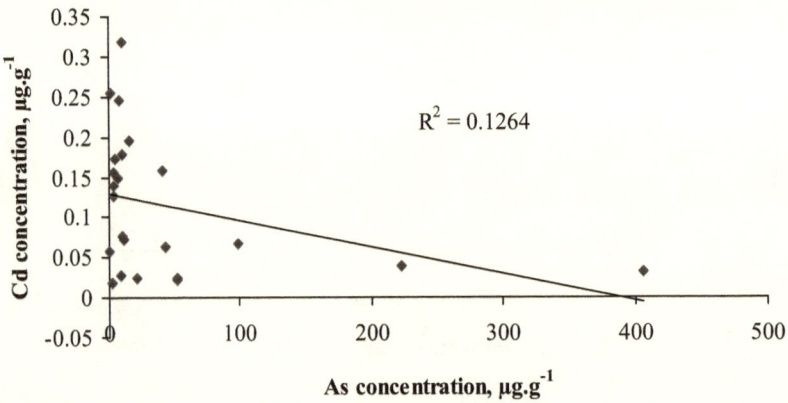

Figure 5.14. Cluster plot of arsenic and cadmium in plant samples.

This study showed that plants can maintain or even increase the concentrations of the essential elements when grown on soil with potentially harmful nutrients like arsenic, cadmium and lead. However, the samples analyzed in this study were mostly leafy parts of the plants. The overall composition of heavy metals in a plant body depends on a number of factors such as the metals and their chemical composition as well as the plant species and population involved.[166]

CHAPTER 6: CONCLUSIONS AND FUTURE WORK

6.1. CONCLUSIONS

The principal goal of this thesis was to develop an efficient extraction method for arsenic in plants that had grown on arsenic contaminated soils. An efficient but simple method of arsenic extraction was lacking. In the development of the extraction process, it was also sought to gain better understanding of the extraction processes, the stability of arsenic species in the extraction processes, the speciation of arsenic in plants, and the influence of solvents and plant matrices on the extraction of arsenic. An efficient extraction method was developed with double the extraction efficiency of current methods. The multi-elemental composition of plants and the relationship of arsenic with the other elements in plants were also investigated. From this study several useful and interesting results were obtained. The major findings of the current study will be briefly discussed below.

6.1.1. Development of a Sequential Extraction Method for Arsenic in Plants

In the development of the sequential extraction method, the efficiencies of solvent-Soxhlet and solvent-sonication methods were compared. A number of plants comprising field horsetail (*Equisetum arvense*), a grass (*Puccinellia sp.*) and smooth horsetail (*Equisetum laevigatum*) were extracted for arsenic. The solvent-sonication method was found to be the more efficient and simpler of the two methods. Among the solvents (water, methanol, 1:1 water-methanol, dilute HCl of varying strengths, 0.3 M-H_3PO_4 and 0.1 M-NaOH) dilute HCl (0.05 M or 0.1 M) was found to be the most efficient for extracting arsenic from plants.

Spikes of arsenite (As(III)), arsenate (As(V)), methylarsonic acid (MA) and dimethylarsinic acid (DMA) were used to determine the recovery and stability using the above solvents. The recovery of spikes conducted individually and mixed together was 90-110% by all solvents and methods, and the species were also found to be stable in all extraction methods employed. The complete recovery of the spikes in all media in contrast to the difficult and incomplete extraction of the arsenic from some plant samples suggested that the arsenic species in real samples were chemically and/or physically bound to the plant matrix.

The plants grown on arsenic contaminated soils contained higher amounts of arsenic the majority of which was inorganic. The trace organoarsenic species were found only in 1:1 water-methanol extracts. The water-alcohol mixture extracted smaller amounts of the inorganic species facilitating the determination of the organoarsenic species. Experiments were conducted with 100% methanol, 50% methanol (1:1 v/v water/methanol) and water (DDW). The results showed that methanol extraction of the inorganic species was poor and extraction of the organoarsenic species was poor in water extracts. All arsenic species (As(III), DMA, MA and As(V)) were detected in 50% methanol. However, DMA was better extracted in 100% methanol.

Simulated gastric fluid extraction (GFE) and water extraction had similar efficiency. However, the advantage of GFE was its direct analysis by ICPAES and HPLC-HGAAS, unlike the 1:1 water-methanol extract that required the removal of organic solvent prior to analysis. Both organic and inorganic arsenic compounds were detected in the GFE extracts. The GFE, however, lacked the general efficiency required for extracting arsenic from the various terrestrial plants. In order to achieve better arsenic extraction efficiency, other enzymatic solutions were explored. A combination of pepsin and salt in 0.1 M-HCl (PHS) was found to be more efficient in extracting arsenic. The PHS extracted significantly more arsenic than that extracted by the sequential method from a number of plants, however, lacked the general efficiency to be applicable to a variety of plants.

Although dilute HCl was efficient in extracting arsenic from plants the determination of organic species was difficult in this media. Since speciation is a critical part of the extraction process of arsenic from plants, the extraction sequence was optimized by using several extraction schemes. The optimization experiments showed that a sequential extraction with 1:1 water-methanol followed by 0.1 M-HCl was most useful in extracting and speciating both organic and inorganic arsenic from plants.

Arsenic in Plants: Extraction and Speciation

The detection limits of the organoarsenic species in the samples were lowered by the process of preconcentration prior to analysis. Further optimization of the method reduced extraction steps from five to three and reduced the sample mass thus improving analysis time and sample consumption. Sample preservation was also achieved by sequential extraction of sample, and time was saved by batch drying of the water-alcohol extracts instead of roto-evaporation of each sample individually. The HPLC-HGAAS was found to be efficient enough to determine all arsenic species extracted from plants by the sequential method.

The developed sequential method was applied to the extraction of arsenic from a number of plants sampled from arsenic impacted soil of Yellowknife in NWT, Canada. Most of these samples were grass species which are usually hard to grind and extract. While 1:1 water-methanol extraction efficiency ranged between 4-89% with a median at about 29%, the efficiencies of the sequential method ranged between 27-128% with a median at 46%. The sequential method extracted overall 100% more arsenic than the traditional water-methanol method. For the Deloro plants extracted using this method, the 1:1 water-methanol process extracted 44% of the overall total arsenic similar to the published results whereas the sequential method extracted 85%. Thus the developed method improves the extraction efficiency to almost double of the traditional water-methanol method.

6.1.2. Arsenic in Plants from Deloro by Sequential Method

Deloro, Ontario was a historical gold mining area and was severely contaminated by arsenic from the mining activities. Over a period spanning 2001-2004, 124 plant samples were collected from Deloro site along with a number of soil samples. A systematic study over three seasons using many plant species was conducted. Total arsenic in all plant and soil samples was determined by ICPAES and low arsenic containing samples were retested by HGAAS. Arsenic in plants was extracted by the sequential method and speciated by HPLC-HGAAS.

The total arsenic in plants of Deloro ranged from 1 to 500 $\mu g.g^{-1}$. Among the species, total arsenic content varied widely. The horsetails

(*Equisetum sp.*) contained the most arsenic with an overall mean 127 µg.g^{-1}. The field horsetail (*E. arvense*) which is valued as a food, medicinal and a silica producing plant accumulated the highest amount of arsenic with a mean of 159 µg.g^{-1}. The Canada goldenrod (*Solidago canadensis*), panicle aster (*Aster lanceolatus*), and purple loosestrife (*Lythrum salicaria*) species contained arsenic on average 5, 8 and 12 µg.g^{-1}, respectively. Another fern, sensitive fern (*Onoclea sensibilis*), contained overall arsenic comparatively higher amounts, 26.4 µg.g^{-1}.

The seasonal variation in the total arsenic content was apparent in the plants. All plants accumulated arsenic up to middle of the season (July) and a few continued accumulating until the end of the season (September). The field horsetail and sensitive fern are the most prominent of the latter category of plants.

The analysis of arsenic in different parts of plants showed that the leaves contained the highest amounts of arsenic. This is significant in view of the fact that in many plants, leaves are the edible parts for both humans and wildlife.

The plants which contained ~ 5 µg.g^{-1} or more total arsenic were extracted by the sequential method. The speciation results showed that As(V) and As(III) were the major, and MA and DMA were the minor (usually trace) species in the plants sampled from Deloro. Among the organoarsenic species, DMA was more abundant in the plants studied. Another advantage of sequential method was in its ability to determine trace organoarsenic species in plants containing high amount of inorganic arsenic.

The ratio of As(V) to As(III) in plants was studied. Results showed that water-methanol extracted significantly higher amount of As(V) than As(III) from all plants. Similar observations were also reported by other researchers. Interestingly, plants contained overall approximately 50/50 As(V)/As(III) in all plant species. When the ratios of As(V)/As(III) were plotted against total arsenic content of a number of plants no correlation was observed between the two parameters. Similar results were observed for the As(V)/As(III) ratio and total extracted arsenic. This showed that the assimilation of arsenic in terms of the ratios of major arsenic species (As(V) and As(III)) in plants was independent of the total arsenic absorbed by the plants.

It was observed that the extraction efficiencies varied widely among the plant species from as low as 13% up to 100%. The reason why some plants were difficult to extract was not clear. However, when the ratios of As(III)/As(V) of the various plant species were plotted against the arsenic extraction efficiencies of the plants a general trend was observed. A significant correlation between higher As(III) to lower extraction efficiency was found. That the As(III) is harder to extract was also supported by the general observation that more As(V) than As(III) was extracted by 1:1 water-methanol. The difficulty of extracting As(III) may lie in the fact that much of the arsenite is chemically and/or physically bound to the plant matrix.

It was observed initially that extraction efficiency was related to the total arsenic content of plants. The earlier results showed that the extraction efficiency decreased as total arsenic content in plants increased. However, the observation was based on extraction efficiencies of arsenic from different plant species. Further investigation of this relationship was carried out using the same plant species thus minimizing the effect of differences found among plant species. The extraction efficiencies of different plant species were plotted against the total arsenic content of the plants. When separated into different plant species clear trends were observed. For the field horsetails, purple loosestrifes, and all horsetails together, the extraction efficiency was independent of total arsenic content of plants. The Canada goldenrods and sensitive ferns showed decreasing extraction efficiency with increasing total arsenic content. On the other hand, an increasing trend of extraction efficiency with increasing total arsenic content was observed for panicle aster. These trends have not been reported previously and clearly indicated matrix effects on extraction efficiency.

6.1.3. Further Aspects of Arsenic Extraction Study

Our earlier results showed that sequential methods extracted on the average 100% more arsenic from plants. Further simplification of the arsenic extraction process was possible by extracting plants in one solvent of 0.1 M-HCl. A comparison of the extraction efficiencies of the methods showed that extraction with 0.1 M-HCl alone was as efficient as the sequential method because the organoarsenic components of plants were very small. However, as discussed above, no organoarsenic species could be detected in 0.1 M-HCl extracts and the sequential method would be useful in that regard.

The determination of trace amounts of organoarsenic species in plants grown on arsenic impacted soils was a challenge due to the presence of larger amounts of the inorganic species. Generally, HPLC-ICPMS has been used for their detection. But, the prohibitive cost of procurement and maintenance prevent many laboratories from using this sophisticated instrument. As an alternative, a less expensive method such as HPLC-HGAAS was desired. The plant extracts in 1:1 water-methanol and 100% methanol were speciated by the two methods and the results compared. The results from two methods agreed very well showing that the use of HPLC-HGAAS is sufficient for this purpose.

In order to compare the species in the original plants, extracts and residues, XANES analysis were conducted to determine the As(III) and As(V) fractions in solid plant samples without extraction, and in the solid residues after extraction. The XANES results indicated the presence of As(III), As(V) and S-bound arsenic (e.g., As-Glu and As-PC) species in plants.

XANES were performed also on freshly frozen plants in order to compare species transformation because of drying and grinding. The different plant species were tested and results showed that the arsenic species essentially remained consistent during washing, drying and grinding processes. This also showed that the As(III)/As(V) pair in plants was stable. Comparison of As(III) and As(V) in the original plant and the residue of extraction showed similarity in most cases, however, some increase in the As(V) fractions in the residues of sensitive fern and purple loosestrife was observed.

The results of arsenic species in the solid samples by XANES and in the extracts by HPLC-HGAAS were compared. A direct comparison of results from the methods was difficult due to highly varied (16-100%) extraction efficiencies of plants and wide difference (percent versus part per million) in the sensitivities of the methods. For a number of plants, As(V) fraction in the HPLC-HGAAS was found to be higher than that in the original sample suggesting some species transformation during the extraction process (5.4). The possible source(s) of the increased As(V) may be either As(III) or As(III)-Glu (and any S-bound As).

Seasonal variation in the arsenic species was observed earlier. The XANES analysis of a number of plants collected in May, July and September showed similar variation that was highly conspicuous in field horsetail plants. The predominance of the As(III) species in purple loosestrife and sensitive fern throughout the season was observed. These two plants were also hard to extract with extraction efficiencies of $(18 \pm 2)\%$ and $(48 \pm 8)\%$, respectively.

Thirty elements, including arsenic, were determined in twenty-two plant samples. The concentrations of arsenic in the plants varied widely. The correlation of arsenic concentration with those of the other elements was investigated. Association between the concentrations of As and S, and As and Ca was observed. A correlation between As and S concentrations was expected since the defense mechanism of plant triggers the formation of arsenic binding complexes, As-PC and As-Glu, in plant body with increased toxic arsenic in the plant system.

6.2. PROPOSALS FOR FUTURE WORK

6.2.1. Effect of Plant Matrix on Arsenic Speciation

It was shown in this study that plant matrices played a role in the extraction of arsenic from plants. However, their role in the speciation of arsenic is not known. The aqueous environment of each plant extract is different from one plant to the other and certainly much different from the pure water solutions of the arsenic standards used for the calibration purposes. In this study, DMA was found to be particularly affected by the matrix difference between standard solution and plant extracts, and even within the standard solutions of DMA and other arsenic species. Further studies involving standard addition and external calibration methods, matrix matching and spiking experiments may reveal interesting knowledge and the explanation for the discrepancy.

6.2.2. Extraction Efficiency and As(III) in Plants

A relationship between low extraction efficiency and As(III) content of plant was shown in this study. Purple loosestrife and sensitive fern contained As(III) more than 90% of total As throughout the season and had low extraction efficiency. Canada goldenrod also showed low extraction efficiency most likely due to its high As(III) content. In contrast, the horsetails early in the season showed high As(III) content yet extraction efficiency of this plant was much higher than the other plants. More similar plants may be identified and extracted to verify the relationship between low EE and high As(III) content of the plants. Studies using XANES (X-ray absorption near edge spectroscopy) for the arsenic species and EXAFS (Extended X-ray Absorption Fine Structure) for their coordination environment in plant matrices may provide important knowledge on arsenic assimilation by the plants.

6.2.3. Field Horsetail as Arsenic Accumulator

The horsetails particularly the field horsetail (*Equisetum arvense*) is used as food, medicine, cosmetics and in technology for the production of biomimetic geolite. In this and other studies, field horsetail in nature was found to accumulate arsenic as high as 700 $\mu g.g^{-1}$. In this study, the highest arsenic extraction efficiency was also accomplished in this plant. Controlled research could be conducted to realize its maximum potential for arsenic accumulation. The field horsetail which, in this study, was found to contain the highest amount of arsenic (> 500 $\mu g.g^{-1}$) also stands straight up from soil (average height about 45 cm) and is easy to harvest. The potential of field horsetail for the phytoremediation of arsenic impacted soils could be investigated. The arsenic accumulating ferns such as the Chinese brake fern (*Pteris vittata L.*) are already being investigated for potential phytoremediation in southern climates. The arsenic contaminated areas such as Deloro may be suitable for similar investigation and application.

6.2.4. Transformation of Inorganic Arsenic during Extraction

Arsenic in Plants: Extraction and Speciation

The stability of the arsenic species (As(III), As(V), MA and DMA) was demonstrated in the extraction processes from the spike recovery by HPLC-HGAAS and in the fresh frozen and dry plants by XANES experiments. A number of plants, however, which had proportionately low amounts of As(V) in the original sample, showed proportionately higher amounts of As(V) in the extracts. In the original sample, these plants were found to contain high amounts of As(III) and As(III)-Glu (i. e., the S-bound species). In order to maintain the mass balance, the source of the increased As(V) in the extracts must be either As(III) or As(III)-Glu or both. Since the spike recovery and XANES studies showed stability of As(III) in solution and in the wet and dry plants, the source of As(V) may be the As(III)-Glu species. For this, the stability of As(III)-Glu complexes may be tested by spiking it in the solvent-sonication extraction processes. However, the above conclusion might turn out to be simplistic and only studies using standard species concerned in the extraction process and direct speciation methods such as the XANES and EXAFS may reveal the kinetic and physical transformations of arsenic species from the plant to the extract.

REFERENCES

[1] Oremland, R.S. and Stolz, J.F., the Ecology of Arsenic, SCIENCE, 300, 2003.

[2] Nriagu, J.O., Arsenic poisoning through the ages, in *Environmental Chemistry of Arsenic*, Frankenberger Jr. W.T., Ed. (Dekker, New York, 2002), pp. 1–26.

[3] Goessler, W., and Kuehnelt, D., Analytical methods for the determination of arsenic and arsenic compounds in the environment, in *Environmental Chemistry of Arsenic*, W. T. Frankenberger Jr., Ed. (Dekker, New York, 2002), pp. 27-50.

[4] Greenwood, N.N. and Earnshaw, A. (1984) Chemistry of the Elements, 1542 p. Pergamon Press, New York.

[5] Internet Source, Arsenic in the environment, Dept. Geology, Univ. Otago, New Zealand, http://www.otago.ac.nz/geology/features/metals/arsenic.html, April 19, 2005.

[6] Cullen, W.R. and Reimer, K.J., (1989), Arsenic speciation in the environment, Chemical Reviews, 89, 713–764.

[7] Matschullat, J., Arsenic in the Geosphere – a review, The Science of the Total Environment 249 (2000) 297-312.

[8] Feldmann, J., Devalla, S., Raab, A. and Hansen, H.R., Analytical Strategies for Arsenic Speciation in Environmental and Biological Samples, in *Organic metal and metalloid species in the environment*, Hirner, A.V. and Emons, H., Eds., (Springer-Verlag, Berlin Heidelberg 2004), pp. 41-70.

[9] Azcue, J.M. and Nriagu, J.O., Arsenic: Historical Perspectives. In *J.O. Nriagu, Ed., Arsenic in the Environment Part 1: Cycling and Characterization*, (Wiley, New York, 1994), p. 430.

[10] Tanaka, T., Distribution of Arsenic in the Natural Environment with an Emphasis on Rocks and Soils, Applied Organometallic Chem., 1988, 2, 283-295.

[11] Francesconi, K.A. and Kuehnelt, D., Arsenic Compounds in the Environment, in *Environmental Chemistry of Arsenic*, W.T. Frankenberger Jr., Ed. (Dekker, New York, 2002), pp. 51-94.

[12] Mandal, B.K and Suzuki, K.T., Arsenic Round the World: a Review, Talanta (2002), 58(1), 201-235.

[13] Magalhaes, M. and Clara, F., Arsenic: An Environmental Problem Limited by Solubility, Pure and Applied Chemistry (2002), 74(10), 1843-1850.

[14] Peters, G.R., McCurdy, R.F. and Hindmarsh, J.T., Environmental Aspects of Arsenic Toxicity, Critical Reviews in Clinical Laboratory Sciences (1996), 33(6), 457-493.

[15] Koch, I., Feldmann, J., Wang, L., Andrewes, P., Reimer, K. J. and Cullen, W.R., Arsenic in the Meager Creek Hot Springs Environment, British Columbia, Canada, Science of the Total Environment

(1999), 236(1-3), 101-117.

[16] Koch, I., Wang, L., Ollson, C.A., Cullen, W.R. and Reimer, K.J., The Predominance of Inorganic Arsenic Species in Plants from Yellowknife, Northwest Territories, Canada, Environmental Science and Technology (2000), 34(1), 22-26.

[17] Alam, M.G.M., Allinson, G., Stagnitti, F. and Tanaka, A.; Westbrooke, M., Arsenic Contamination in Bangladesh Groundwater: a Major Environmental and Social Disaster, International Journal of Environmental Health Research (2002), 12(3), 235-253.

[18] Hadi, A. and Parveen, R., Arsenicosis in Bangladesh: Prevalence and Socio-Economic Correlates, Public Health (2004), 118(8), 559-64.

[19] Eisler, R., Arsenic Hazards to Humans, Plants, and Animals from Gold Mining, Reviews of Environmental Contamination and Toxicology (2004), 180, 133-65.

[20] Alam, A.M.S., Islam, M.A., Rahman, M.A., Ahmed, E., Islam, S., Sultana, K.S. and Siddique, M.N., Transport of Toxic Metals through the Major River Systems of Bangladesh, Journal of the Chemical Society of Pakistan (2004), 26(3), 328-332.

[21] Ridgway, J., Breward, N., Langston, W. J., Lister, R., Rees, J. G. and Rowlatt, S. M., Distinguishing between Natural and Anthropogenic Sources of Metals Entering the Irish Sea, Applied Geochemistry (2003), 18(2), 283-309.

[22] Liu, E., Shen, J., Zhu, Y., Xia, W. and Zhu, G., Source Analysis of Heavy Metals in Surface Sediments of Lake Taihu, Hupo Kexue (2004), 16(2), 113-119. (2005 ACS on SciFinder ®)

[23] Kamal, A.S.M., and Parkpian, P., Arsenic Contamination in Hizla, Bangladesh: Sources, Effects and Remedies, Science Asia (2002), 28(2), 181-189.

[24] Gong, Z., Lu, X., Ma, M., Watt, M.C. and Le, X.C., Arsenic Speciation Analysis, Talanta (2002), 58, 77-96.

[25] Koch, I., Ollson, C.A., Potten, J. and Reimer, K.J., Arsenic In Vegetables: An Evaluation Of Risk From The Consumption of Produce from Residential and Mine Gardens in Yellowknife, Northwest Territories, Canada, Reviews in Food and Nutrition Toxicity (2003), 116-140.

[26] B'Hymer, C. and Caruso, J.A., Arsenic and its Speciation Analysis Using High-Performance Liquid Chromatography and Inductively Coupled Plasma Mass Spectrometry, Journal of Chromatography, A (2004), 1045(1-2), 1-13.

[27] Francesconi, K.A. and Kuehnelt, D., Determination of Arsenic Species: A Critical Review of Methods and Applications, 2000–2003, Analyst (2004), 129, 373-395.

[28] Jin, Y., Sun, G., Li, X., Li, G., Lu, C. and Qu, L., Study on the Toxic Effects Induced by Different

Arsenicals in Primary Cultured Rat Astroglia, Toxicology and Applied Pharmacology (2004), 196(3), 396-403.

[29] Xie, Y., Trouba, K., Liu, J., Waalkes, M.P. and Germolec, D.R., Biokinetics and Subchronic Toxic Effects of Oral Arsenite, Arsenate, Monomethylarsonic Acid, and Dimethylarsinic Acid in V-Ha-Ras Transgenic (Tg.AC) Mice, Environmental Health Perspectives (2004), 112(12), 1255-1263.

[30] Vasken, A.H., Zakharyan R.A., Avram, M.D, Kopplin, M.J. and Wollenberg, M.L., Oxidation and Detoxification of Trivalent Arsenic Species, Toxicology and applied pharmacology (2003), 193(1), 1-8.

[31] Thomas, David J.; Styblo, Miroslav; Lin, Shan, the Cellular Metabolism and Systemic Toxicity of Arsenic, Toxicology and Applied Pharmacology (2001), 176(2), 127-144.

[32] Petrick, J.S., Ayala-Fierro, F., Cullen, W.R., Carter, D.E. and Vasken A.H., Monomethylarsonous Acid (MMAIII) is More Toxic than Arsenite in Chang Human Hepatocytes, Toxicology and Applied Pharmacology (2000), 163(2), 203-207.

[33] Knauer, K., Behra, R., and Hemond, H., Toxicity of Inorganic and Methylated Arsenic to Algal Communities from Lakes along an Arsenic Contamination Gradient, Aquatic Toxicology (1999), 46(3-4), 221-230.

[34] Shi, H., Hudson, L.G., Ding, W., Wang, S., Cooper, K.L., Liu, S., Chen, Y., Shi, X., and Liu, K.J., Arsenite Causes DNA Damage in Keratinocytes Via Generation of Hydroxyl Radicals, Chemical Research in Toxicology (2004), 17(7), 871-878.

[35] Rodriguez, V.M., Jimenez-Capdeville, M.E. and Giordano, M., The Effects of Arsenic Exposure on the Nervous System, Toxicology Letters (2003), 145(1), 1-18.

[36] Ratnaike, R.N., Acute and chronic arsenic toxicity, Postgraduate Medical Journal (2003), 79(933), 391-396.

[37] Tchounwou, P.B., Centeno, J.A. and Patlolla, A.K., Arsenic toxicity, mutagenesis and carcinogenesis: a health risk assessment and management approach, Molecular and cellular biochemistry (2004), 255(1-2), 47-55.

[38] Ng, J.C., Wang, J. and Shraim, A., A Global Health Problem Caused by Arsenic from Natural Sources, Chemosphere (2003), 52(9),1353-1359.

[39] Nordstrom, D.K., An overview of Arsenic Mass-Poisoning in Bangladesh and West Bengal, India, Editor(s): Young, C., Minor Elements 2000: Processing and Environmental Aspects of As, Sb, Se, Te, and Bi, [Symposium], Salt Lake City, UT, United States, 2000 (2000), Meeting Date 2000, 21-30.

[40] Tondel, M., Rahman, M., Magnuson, A., Chowdhury, I.A.; Faruquee, M.H., and Ahmad, S.A., The

Relationship of Arsenic Levels in Ddrinking Water and the Prevalence Rate of Skin Lesions in Bangladesh, Environmental Health Perspectives (1999),107(9), 727-729.

[41] Milton, A.H., Hasan, Z., Rahman, A. and Rahman, M., Chronic Arsenic Poisoning and Respiratory Effects in Bangladesh, Journal of Occupational Health (2001), 43(3), 136-140.

[42] Rahman, M., Tondel, M., Ahmad, S.A., Chowdhury, I.A., Faruquee, M.H., and Axelson, O., Hypertension and Arsenic Exposure in Bangladesh, Hypertension (1999), 33(1), 74-78.

[43] Chen, Y., and Ahsan, H., Cancer Burden from Arsenic in Drinking Water in Bangladesh, American journal of public health (2004), 94(5), 741-4.

[44] Rahman, M., Tondel, M., Ahmad, S.A. and Axelson, O., Diabetes Mellitus Associated with Arsenic Exposure In Bangladesh, American journal of epidemiology (1998), 148(2),198-203.

[45] Tseng, C.H., Tseng, C.P., Chiou, H.Y., Hsueh, Y.M., Chong, C.K. and Chen, C.J., Epidemiologic Evidence of Diabetogenic Effect of Arsenic, Toxicology letters (2002), 133(1), 69-76.

[46] Hindmarsh, T.J., Trace Elements in Human Health and Disease: an Update, Arsenic: its Clinical and Environmental Significance, 2000, 13 (1), 165-162.

[47] Dabeka, R.W., McKenzie, A.D., Lacroix, Gladys, M.A.; Cleroux, C., Bowe, S., Graham, R.A., Conacher, H.B.S. and Verdier, P., Survey of Arsenic in Total Diet Food Composites and Estimation of the Dietary Intake of Arsenic by Canadian Adults and Children, Journal of AOAC International (1993), 76(1), 14-25.

[48] Meharg, A.A. and Rahman, M.M., Arsenic Contamination of Bangladesh Paddy Field Soils: Implications for Rice Contribution to Arsenic Consumption, Environmental Science and Technology (2003), 37(2), 229-234.

[49] Roychowdhury, T., Uchino, T., Tokunaga, H. and Ando, M., Survey of Arsenic in Food Composites From an Arsenic-Affected Area of West Bengal, India, Food and Chemical Toxicology (2002), 40(11),1611-1621.

[50] Das, H.K., Mitra, A.K., Sengupta, P.K., Hossain, A., Islam, F. and Rabbani, G.H., Arsenic Concentrations in Rice, Vegetables, and Fish in Bangladesh: A Preliminary Study, Environment International (2004), 30(3), 383-387.

[51] Patra, M., Bhowmik, N., Bandopadhyay, B. and Sharma, A., Comparison of Mercury, Lead and Arsenic with Respect to Genotoxic Effects on Plant Systems and the Development of Genetic Tolerance, Environmental and Experimental Botany (2004), 52(3), 99-223.

[52] Jha, A.B. and Dubey, Carbohydrate Metabolism in Growing Rice Seedlings under Arsenic Toxicity, R.S., Journal of Plant Physiology (2004), 161(7), 867-872.

[53] Albert, W.B. and Arndt, C.H., Concentration of soluble arsenic as an index of arsenic toxicity to plants, S. Car. Agr. Expt. Sta., Ann. Rept., (1931), 47-8.

[54] Clements, H.F.; Heggeness, H.G., Arsenic toxicity to plants, Hawaii Agr. Expt. Sta., Rept., (1940), Volume Date 1939, 77-8.

[55] Inskeep, W.P., McDermott, T.R. and Fendorf, S., Arsenic (V)/(III) cycling in soils and natural waters; Chemical and Microbiological processes in *Environmental Chemistry of Arsenic*, W.T. Frankenberger Jr., Ed. (Dekker, New York, 2002), pp. 183-216.

[56] Kempton, J.H., Lindberg, R.D. and Runnells, D.D., Numerical Modeling of Platinum Eh Measurements by Using Heterogeneous Electron Transfer Kinetics, In *Chemical Modeling in Aqueous Systems II*, Chapter 27, eds., Melchior, D.C. and Bassett, R.L., ACS Symposium Series 416, (1990), ACS, WA, D.C. pp 339-349.

[57] The Deloro Mine Site Cleanup Project and Draft Cleanup Plan, Summary from internet at http://www.ene.gov.on.ca/envision/deloro/index.htm, March 03, 2005.

[58] Szpunar, J., Bouyssiere, B. and Lobinski, R., Advances in Analytical Methods for Speciation of Trace Elements in the Environment, in Organic Metal and Metalloid Species in the Environment, Hirner, A.V. and Emons, H., Eds., (Springer-Verlag, Berlin Heidelberg 2004), pp. 17-40.

[59] Mahin, E.G., The Determination of Total Arsenic Acid in London Purple, Purdue Univ., Journal of the American Chemical Society (1907), 28, 1598-1601.

[60] Webb, S.M., Gaillard, J.F., Ma, L.Q. and Tu, C., XAS Speciation of Arsenic in a Hyper-Accumulating Fern, Environ. Sci. Technol. (2003), 37, 754-760.

[61] Smith, P.G., Koch, I., Gordon, R.A., Mandoli, D.F., Chapman, B.D. and Reimer, K.J., X-ray Absorption Near-Edge Structure Analysis of Arsenic Species for Application to Biological Environmental Samples, Environ. Sci. Technol. (2005), 39, 248-254.

[62] Langdon, C., Mehrag, A.A., Feldmann, J., Balger, T., Charnock, J., Farquhar, M., Piearce, T., Semple, K. and Cotter-Howells, J., Arsenic-Speciation in Arsenate-Resistant and Non-Resistant Populations of the Earthworms, Lumbricus rubellus, J. Environ. Monit., 2002, 4, 603-608.

[63] Rodushkin, I., Ruth, T. and Huhtasaari, A., Comparison of Two Digestion Methods for Elemental Determinations in Plant Material by ICP Techniques, Analytica Chimica Acta, (1999), 378(1-3),191-200.

[64] Arunachalam, J., Mohl, C., Ostapczuk, P. and Emons, H., Multielement Characterization of Soil Samples with ICP-MS for Environment Studies, Fresenius' Journal of Analytical Chemistry (1995), 352(6), 577-81.

[65] Elkhatib, E.A., Bennett, O.L. and Wright, R.J., Determination of Total Arsenic in Soil by Differential Pulse Polarography, Soil Science Society of America Journal (1983), 47(4), 836-8.

[66] Prakash, R., Srivastava, R.C. and Seth, P.K., Direct Estimation of Total Arsenic Using a Novel Metal Side Disk Rotating Electrode, Electroanalysis (2003),15(17),1410-1414.

[67] Tang, S., Zhou, M. and Tong, X., Determination of Total Arsenic in Soil by Nondispersive Hydride-Generation Atomic Fluorescence Spectrometry, Fudan Xuebao, Ziran Kexueban (1985), 24(3), 271-7. (Copyright 2005 ACS on SciFinder (R))

[68] Jimenez De Blas, O., Mateos, N.R., and Sanchez, A.G., Determination of Total Arsenic and Selenium in Soils and Plants by Atomic Absorption Spectrometry with Hydride Generation and Flow Injection Analysis Coupled Techniques, Journal of AOAC International (1996), 79(3), 764-768.

[69] Zhu, B. and Tabatabai, M.A., An Alkaline Oxidation Method for Determination of Total Arsenic and Selenium in Sewage Sludges, Journal of Environmental Quality (1995), 24(4), 622-6.

[70] Narukawa, T., Yoshimura, W. and Uzawa, A., Determination of Total Arsenic in Environmental and Geological Samples by Electrothermal Atomic Absorption Spectrometry Using a Tungsten Furnace after Solvent Extraction and Cobalt(III) Oxide Collection, Bulletin of the Chemical Society of Japan (1999), 72(4), 701-706.

[71] Bamford, S.A., Wegrzynek, D., Chinea-Cano, E. and Markowicz, A., Application of X-Ray Fluorescence Techniques for the Determination of Hazardous and Essential Trace Elements in Environmental and Biological Materials, Nukleonika (2004), 49(3), 87-95.

[72] Schwenke, H., Beaven, P.A. and Knoth, J., Application of total reflection x-ray fluorescence spectrometry in trace element and surface analysis, Fresenius' Journal of Analytical Chemistry (1999), 365(1-3), 19-27.

[73] Mukhtar, S., Haswell, S.J., Ellis, A.T. and Hawke, D.T., Application of Total-Reflection X-Ray Fluorescence Spectrometry to Elemental Determinations in Water, Soil and Sewage Sludge Samples, Analyst (1991), 116(4), 333-8.

[74] Krachler, M., Hendrik, E., Barbante, C., Cozzi, G., Cescon, P., and Shotyk, W., Inter-method comparison for the determination of antimony and arsenic in peat samples, Analytica Chimica Acta, 2002, 458, 387-396.

[75] Beccaloni, E., Musmeci, L. and Stacul, E, Determination of arsenic in environmental solid matrix, Anal. Bioanal Chem., 2002, 374, 1230-1236.

[76] Burguera, M. and Burguera, J.L., Analytical Methodology for Speciation of Arsenic in Environmental and Biological Samples, Talanta (1997), 44(9), 1581-1604.

[77] Schmeisser, E., Goessler, W., Kienzl, N. and Francesconi, K.A., Volatile Analytes Formed from Arsenosugars: Determination by HPLC-HG-ICPMS and Implications for Arsenic Speciation Analyses, Anal. Chem. 2004, 76, 418-423.

[78] Braman, R.S., Justen, L.L. and Foreback, C.C., Direct Volatilization-Spectral Emission Type Detection System for Nanogram Amounts Of Arsenic and Antimony, Analytical Chemistry (1972), 44(13), 2195-9.

[79] Kohlmeyer, U., Kuballa, J. and Jantzen, E., Simultaneous Separation of 17 Inorganic and Organic Arsenic Compounds in Marine Biota By Means Of High-Performance Liquid Chromatography/Inductively Coupled Plasma Mass Spectrometry, Rapid Commun. Mass Spectrom., 2002, 16, 965-974.

[80] Sloth, J.J., Larsen, E.H. and Julshamn, K., Determination of Organoarsenic Species in Marine Samples Using Gradient Elution Cation Exchange HPLC-ICP-MS, Journal of Analytical Atomic Spectrometry (2003), 18(5), 452-459.

[81] Wangkarn, S. and Pergantis, S.A., High-Speed Separation of Arsenic Compounds Using Narrow-Bore High-Performance Liquid Chromatography On-Line With Inductively Coupled Plasma Mass Spectrometry, Journal of Analytical Atomic Spectrometry (2000), 15(6), 627-633.

[82] Sun, Y.C.; Lee, Y.S.; Shiah, T.L.; Lee, P.L.; Tseng, W.C. and Yang, M.H., Comparative Study on Conventional and Low-Flow Nebulizers for Arsenic Speciation by Means of Microbore Liquid Chromatography with Inductively Coupled Plasma Mass Spectrometry, Journal of Chromatography, A (2003), 1005(1-2), 207-213.

[83] Bohari, Y., Lobos, G., Pinochet, H. and Pannier, F., Astruc, A. and Potin-Gautier, M., Speciation of Arsenic in Plants by HPLC-HG-AFS: Extraction Optimization on CRM Materials and Application to Cultivated Samples, J. Environ. Monit., 2002, 4, 596.

[84] Porter, E.K. and Peterson, P.J., Arsenic Accumulation by Plants on Mine Waste (United Kingdom), Stevens Report, Stevens Institute of Technology (1975), 4(4),365-71.

[85] Meharg, A.A. and Hartley-Whitaker, J., Tansley review no. 133, Arsenic Uptake and Metabolism in Arsenic Resistant and Nonresistant Plant Species, New Phytologist (2002), 154(1), 29-43.

[86] Cobb, G.P., Sands, K., Waters, M., Wixson, B.G. and Dorward-King, E., Accumulation of Heavy Metals by Vegetables Grown in Mine Wastes, Environmental Toxicology and Chemistry (2000), 19(3), 600-607.

[87] Ma, L.Q., Komart, K.M., Tu, C., Zhang, W., Cai, Y., Kennelly, E.D., 2001, A Fern that Hyperaccumulates Arsenic, Nature 409, 579.

[88] Tu, C., Ma, L.Q. and Bondada, B., Arsenic Accumulation in the Hyperaccumulator Chinese Brake and Its Utilization Potential for Phytoremediation, Journal of Environmental Quality (2002), 31(5), 1671-1675.

[89] Wei, C.Y. and Chen, T.B., The Ecological and Chemical Characteristics of Plants in the Areas of High Arsenic Levels, Zhiwu Shengtai Xuebao (2002), 26(6), 695-700. (Copyright 2004 ACS on SciFinder (R))

[90] Font, R., Del Rio, M., Velez, D., Montoro, R. and De Haro, A., Use of Near-Infrared Spectroscopy for Determining the Total Arsenic Content in Prostrate Amaranth, Science of the Total Environment (2004), 327(1-3), 93-104.

[91] Pascual Buigues, J. and Cortazar Castilla, M.I., Application of Preconcentration Methodologies to the Determination of Arsenic (III/V) and Total Arsenic in Peach Tree Ash Material Using X-Ray Fluorescence, Agrochimica (1995), 39(2-3),161-168. (Copyright 2005 ACS on SciFinder (R))

[92] Koch, I., Wang, L., Reimer, K.J. and Cullen, W.R., Arsenic Species in Terrestrial Fungi and Lichens from Yellowknife, NWT, Canada, Applied Organometallic Chemistry (2000), 14(5), 245-252.

[93] Geiszinger, A., Goessler, W. and Kosmus, W., Organoarsenic Compounds in Plants and Soil on Top of an Ore Vein, Appl. Organometal. Chem., 2002, 16, 245-249.

[94] Koch, I., Hough, C., Mousseau, S., Mir, K., Rutter, A., Olson, C., Lee, E., Andrewes, P., Granhchino, S., Cullen, B. and Reimer, K., Sample Extraction for Arsenic Speciation, Canadian J. Anal. Sci. Spectros., 2002, 47 (4), 109.

[95] He, B., Fang, Y., Jiang, G. and Ni, Z., Optimization of the Extraction for the Determination of Arsenic Species in Plant Materials by High-Performance Liquid Chromatography Coupled with Hydride Generation Atomic Fluorescence Spectrometry, Spectrochimica Acta, Part B: Atomic Spectroscopy (2002), 57B (11), 1705-1711.

[96] Caruso, J.A., Heitkemper, D.T. and B'Hymer, C., An Evaluation of Extraction Techniques for Arsenic Species from Dreeze-Dried Apple Samples, Analyst, 2001, 126 (2), 136.

[97] Krachler, M. and Emons, H., Extraction of Antimony and Arsenic From Fresh and Freeze-Dried Plant Samples as Determined by HG-AAS, Fresenius J. Anal. Chem., 2000, 368, 702.

[98] Rasul, S.B., Hossain, Z., Munir, A.K.M., Alauddin, M., Khan, A.H. and Hussam, A. Electrochemical Measurement and Speciation of Inorganic Arsenic in Groundwater of Bangladesh, Talanta, 58(1), 33-43, (2002).

[99] Pyles, R.A. and Woolson, E. A., Quantitation and Characterization of Arsenic Compounds in Vegetables Grown in Arsenic Treated Soil, J. Agric. Food Chem., 1982, 30, 866.

[100] Gomez-Ariza, J.L., Sanchez-Rodes, D., Giraldez, I. and Morales, E., Comparison of Biota Sample Pretreatments for Arsenic Speciation with Coupled HPLC-HG-ICP-MS. Analyst, 2000, 125, 401.

[101] Brisbin, J.A. and Caruso, J.A., Comparison of Extraction Procedures for The Determination of Arsenic and Other Elements in Lobster Tissue by Inductively Coupled Plasma Mass Spectrometry, Analyst, 2002, 127 (7), 921.

[102] Zheng, J., Hintelmann, H., Dimock, B. and Dzurko, M.S., Speciation of Arsenic in Water, Sediment, and Plants of The Moira Watershed, Canada, Using HPLC Coupled To High Resolution ICP-MS, Analytical and Bioanalytical Chemistry (2003), 377(1), 14-24.

[103] Heitkemper, D.T., Vela, N.P., Stewart, K.R. and Westphal, C.S., Determination of Total and Speciated Arsenic in Rice by Ion Chromatography and Inductively Coupled Plasma Mass Spectrometry, J. Anal. Atom. Spectrom., 2001, 16, 299.

[104] Montilla, A., in *Sample Treatment for Arsenic Speciation*, Master's Thesis, 2000, Univ. Alberta, Canada.

[105] Vela, N.P., Heitkemper, D.T. and Stewart, K.R., Arsenic Extraction and Speciation in Carrots using Accelerated Solvent Extraction, Liquid Chromatography and Plasma Mass Spectrometry, Analyst, 2001, 126, 1011.

[106] Schmidt, A.C., Reisser, W., Mattusch, P.P. and Wennrich, R., Evaluation of Extraction Procedures for the Ion Chromatographic Determination of Arsenic Species in Plant Materials, Journal of Chromatography A, 2000, 889, 83-91.

[107] Helgensen, H. and Larsen, E. H., Bioavailability and Speciation of Arsenic in Carrots Grown in Contaminated Soil, Analyst, 1998, 123, 791.

[108] Daus, J. Mattusch, R. Wennrich and H. Weiss, Investigation on stability and preservation of arsenic species in iron rich water samples, Talanta (2002), 58, 57.

[109] Ruby, M.V., Schoof, R., Brattin, W., Goldade, M., Post, G., Harnois, M., Mosby, D.E., Casteel, S.W., Berti, W., Carpenter, M., Edwards, D., Cragin, D. and Chappell, W., Advances in Evaluating the Oral Bioavailability of Inorganics in Soil for Use in Human Health Risk Assessment, Environ. Sci. Technol. (1999), 33(21), 3697-3705.

[110] Rodriguez, R.R., Basta, N.T., Casteel, S.W. and Pace, L.W., An in Vitro Gastrointestinal Method to Estimate Bioavailable Arsenic in Contaminated Soils and Solid Media, Environmental Science and Technology (1999), 33(4), 642-649.

[111] Ruby, M.V., Davis, A., Schoof, R., Eberle, S. and Sellstone, C., Estimation of Lead and Arsenic Bioavailability Using a Physiologically Based Extraction Test, Environmental Science and Technology

(1996), 30(2), 422-30.

[112] Schroder, J.L., Basta, N.T., Si, J., Casteel, S.W., Evans, T. and Payton, M., In Vitro Gastrointestinal Method to Estimate Relative Bioavailable Cadmium in Contaminated Soil, Environmental Science and Technology (2003), 37(7), 1365-1370.

[113] Hamel, S.C., Ellickson, K.M., Lioy, P. J., The Estimation of The Bioaccessibility of Heavy Metals in Soils Using Artificial Biofluids by Two Novel Methods: Mass-Balance and Soil Recapture, Science of the Total Environment (1999), 243/244, 273-283.

[114] Gaudard, F., The Silica Content of Some Medicinal Plants, Pharmaceutical Journal (1929), 123, 561.

[115] Waentig, P., the Solubility of the Silicic Acid from Silicic Acid-Containing Medicinal Plants, Hippocrates (1939), 10, 969-71. From: Chem. Zentr., 1940, I, 247.

[116] Mori, M., Suzuki, K., Fuzishiro, K. and Hozumi, M., Component Changes in the Growth of Equisetum arvense, Sagami Joshi Daigaku Kiyo, Shizenkei (1996), 60B, 85-91.

[117] Okada, Y. and Hirai, S., Trace Elements in Edible Plants, Musashi Kogyo Daigaku Genshiryoku Kenkyusho Kenkyu Shoho (1985), Volume Date 1984, (9), 142-6.

[118] Vilarem, G., Perineau, F. and Gaset, A., Exploitation of the Molecular Potential of Plants: Equisetum arvense (Equisetaceae), Economic Botany (1992), 46(4), 401-7.

[119] Valtchev, V.P., Smaihi, M., Faust, A.C., Vidal, L., Equisetum arvense Templating of Zeolite Beta Macrostructures with Hierarchical Porosity, Chemistry of Materials (2004), 16(7),1350-1355.

[120] Brooks, R.R., Holzbecher, J. and Ryan, D.E., Horsetails (Equisetum) as Indirect Indicators of Gold Mineralization, Journal of Geochemical Exploration (1981), 16(1), 21-26.

[121] Wong, H.K.T., Gauthier, A. and Nriagu, J.O., Dispersion and Toxicity of Metals from Abandoned Gold Mine Tailings at Goldenville, Nova Scotia, Canada, Science of the Total Environment (1999), 228(1), 35-47.

[122] Heald, S.M., Brewe, D.L., Stern, E.A., Kim, K.H., Brown, F.T., Jiang, D.T., Crozier, E. D. and Gordon, R.A., XAFS and Micro-XAFS at the PNC-CAT Beamlines, J. Synchrotron Radiat. 1999, 6, 347.

[123] Jiang, D.T. and Crozier, E.D., Glancing -angle XAFS of buried Ultrathin Films, Can. J. Phys. 1998, 76, 621.

[124] Kraft, S., Stumpel, J., Becker, P., Kuetgens, U., High-Resolution X-ray Absorption Spectroscopy with Absolute Energy Calibration for the Determination of Absorption Edge Energies, Rev. Sci. Instrum. 1996, 67, 681.

[125] Ressler, T., WinXAS: A New Software Package Not Only For The Analysis Of Energy-Dispersive XAS Data, J. Phys. IV 1997, 7, C2- 269.

[126] Capelo, J.L., Lavilla, I. and Bendicho, C., Ultrasonic Extraction Followed by Sonolysis-Ozonolysis as a Sample Pretreatment Method for Determination of Reactive Arsenic toward Sodium Tetrahydroborate by Flow Injection-Hydride Generation AAS, Anal. Chem., 2001, 73, 3732.

[127] Munoz, O., Devesa, V., Suner, M. A., Velez, D., Montoro, R., Urieta, I., Macho, M.L. and Jalon, M., Total and Inorganic Arsenic in Fresh and Processed Fish Products, J. Agri. Food Chem., 2000, 48, 4369-4376.

[128] Pickering, I.J., Prince, R.C., George, M.J., Smith, R.D., George, G.N. and Salt, D.E., Reduction and Co-Ordination of Arsenic in Indian Mustard, Plant Physiol (2000), 122: 1171–1177.

[129] Raab, A., Feldmann, J., Meharg, A.A., The Nature of Arsenic-Phytochelatin Complexes in Holcus Lanatus and Pteris Cretica, Plant Physiology, 2004, 134, 1113-1122.

[130] Sneller, F.E.C., van Heerwaarden, L.M., Kraaijeveld-Smit, F.J.L., Ten Bookum, W,M., Koevoets, P.L.M., Schat H. and Verkleij, J.A.C., Toxicity of Arsenate in *Silene Vulgaris*, Accumulation and Degradation of Arsenate-Induced Phytochelatins, New Phytol (1999), 144: 223–232.

[131] Schmoger, M.E.V., Oven, M. and Grill, E., Detoxification of Arsenic by Phytochelatins in Plants, Plant Physiology, 2000, 122, 793

[132] Schoof, R.A., Yost, L.J., Eickhoff, J., Crecelius, E.A., Cragin, D.W., Meacher, D.M. and Menzel, D.B., A Market Basket Survey of Inorganic Arsenic in Food, Food Chem. Toxicol., 1999, 37, 839.

[133] Quaghebeur, M., Rengel, Z. and Smirk, M., Arsenic Speciation in Terrestrial Plant Material using Microwave-Assisted Extraction, Ion Chromatography and Inductively Coupled Plasma Mass Spectrometry, J. Anal. Atom. Spectrom., (2003), 18(2), 128-134.

[134] Mitchell, R.L., Evaluation of Washing Techniques for The Removal of External Fluoride from Ironbark and Grape Leaves, Journal of the Australian Institute of Agricultural Science (1986), 52(2), 99-101.

[135] Worley, R.E., Pecan Leaf Analysis is Not Improved by Washing, HortScience (1993), 28(9), 956.

[136] McCrimmon, J.N., Comparison of Washed and Unwashed Plant Tissue Samples Utilized to Monitor the Nutrient Status of Creeping Bentgrass Putting Greens, Communications in Soil Science and Plant Analysis (1994), 25(7-8), 967-88.

[137] Milstein, L.S., Essader, A., Murrell, C., Pellizzari, E.D., Fernando, R.A., Raymer, J.H. and Akinbo, O., Sample Preparation, Extraction Efficiency, and Determination of Six Arsenic Species Present in Food Composites, J. Agric. Food Chem. (2003), 51, 4180–4184.

[138] Skoog, D. A. and Leary, J. J.Principles of Instrumental Analysis (Fourth Ed.), Saunders College Publishing.

[139] Gallagher, P.A., Wei, X., Shoemker, J.A., Brockhoff, C.A. and Creed, J.T., Detection of Arsenosugars from Kelp Extracts Via IC-Electrospray Ionization-MS-MS And IC Membrane Hydride Generation ICP-MS, J. Anal. At. Spectrom., 1999, 14,1829.

[140] Malik, A.K., Gomez, M., Camara, C., Riepe, H.G. and Bettmer, J., On-line Chloride Interference Removal for Arsenic Determination in Waste Water and Urine by ICP-MS Using a Modified Capillary, International Journal of Environmental Analytical Chemistry (2002), 82(11-12), 795-804.

[141] Niedzielski, P., Siepak, M., and Novotny, K., Determination of Inorganic Arsenic Species As(III) and As(V) by High Performance Liquid Chromatography with Hydride Generation Atomic Absorption Spectrometry Detection, Central European Journal of Chemistry (2004), 2(1), 82-90.

[142] Williamsonia, J., and Eklundia, H., Occurrence of Silica in Grasses, Finland, Paperi ja Puu (1996), 78(10), 597-604.

[143] Bunzl, K., Trautmannsheimer, M., Schramel, P. and Reifenhauser, W., Availability of Arsenic, Copper, Lead, Thallium, And Zinc to Various Vegetables Grown in Slag-Contaminated Soils, Journal of Environmental Quality (2001), 30(3), 934-939.

[144] Schmidt, A.C., Mattusch, J., Reisser, W. and Wennrich, R., Uptake and Accumulation Behaviour of Angiosperms Irrigated with Solutions of Different Arsenic Species, Chemosphere (2004), 56(3), 305-313.

[145] Tyler, G. and Olsson, T., Plant Uptake of Major and Minor Mineral Elements as Influenced by Soil Acidity and Liming, Plant and Soil (2001), 230(2), 307-321.

[146] Marin, A.R., Masscheleyn, P.H. and Patrick, W.J., Jr., Soil Redox-Ph Stability of Arsenic Species and its Influence on Arsenic Uptake by Rice, Plant and Soil (1993), 152(2), 245-53.

[147] Reimann, C., Koller, F., Frengstad, B., Kashulina, G., Niskavaara, H. and Englmaier, P., Comparison of the Element Composition in Several Plant Species and Their Substrate from a 1500000-Km2 Area in Northern Europe, Science of the Total Environment (2001), 278(1-3), 87-112.

[148] Meharg, A.A., Variation in Arsenic Accumulation - Hyperaccumulation in Ferns and Their Allies, New Phytologist (2003), 157(1), 25-31.

[149] Ye, Z.H., Lin, Z.Q., Whiting, S.N., de Souza, M.P. and Terry, N., Possible Use of Constructed Wetland to Remove Selenocyanate, Arsenic, and Boron from Electric Utility Wastewater, Chemosphere (2003), 52(9), 1571-1579.

[150] Jung, M.C., Thornton, I. and Chon, H.T., Arsenic, Sb and Bi Contamination of Soils, Plants, Waters

and Sediments in the Vicinity of the Dalsung Cu-W Mine in Korea, Science of the Total Environment (2002), 295(1-3), 81-89.

[151] Pickering, I.J., Prince, R.C., George, M.J. et al., Reduction and Coordination of Arsenic in Indian mustard, Plant Physiology, 2000, 122: 171—178.

[152] (a) Kapustka, L.A., Lipton, J., GAlbraith, H., Cacela, D. and LeJeune, K., Metal and Arsenic Impacts to Soils, Vegetation Communities and Wildlife Habitat in Southwest Montana Uplands Contaminated By Smelter Emissions: II. Laboratory Phytotoxicity Studies, Environmental Toxicology and Chemistry (1995), 14(11), 1905-1912. (b) Galbraith, H., LeJeune, K. and Lipton, J., Metal and arsenic impacts to soils, vegetation communities and wildlife habitat in southwest Montana uplands contaminated by smelter emissions: I. Field evaluation, Environmental Toxicology and Chemistry (1995), 14(11), 1895-1903.

[153] Kitchin, K.T., Recent Advances in Arsenic Carcinogenesis: Modes of Action, Animal Model Systems, and Methylated Arsenic Metabolites, Toxicology and Applied Pharmacology (2001), 172(3), 249-261.

[154] Hough, C.M.A., Characterization of Arsenic in Short Terrestrial Food Chain Yellowknife, Northwest Territories, M.Sc. Thesis, November 2001, Royal Military College of Canada, Kingston, ON, Ca.

[155] Quaghebeur, M. and Rengel, Z., The Distribution of Arsenate and Arsenite in Shoots and Roots of Holcus Lanatus is Influenced by Arsenic Tolerance and Arsenate and Phosphate Supply, Plant Physiology (2003),132(3),1600-1609.

[156] Viehoever, A. and Prusky, S.C., Biochemistry of Silica, American Journal of Pharmacy (1835-1936) (1938), 110, 99-120.

[157] Calomme, M., Cos, P., D'Haese, P., Vingerhoets, R., Lamberts, L., De Broe, M., Van Hoorebeke, C. and Vaanden Berghe, D., Silicon absorption from stabilized orthosilicic acid and other supplements in healthy subjects, Editor(s): Roussel, A.M., Anderson, R.A. and Favier, A.E., Trace Elements in Man and Animals 10, [Proceedings of the International Symposium on Trace Elements in Man and Animals], 10th, Evian, France, May 2-7, 1999 (2000), Meeting Date 1999,1111-1114.

[158] Carnat, A., Petitjean-Freytet, C., Muller, D. and Lamaison, J.L., Content of Major Constituents of Horsetails, Equisetum arvense L., Plantes Medicinales et Phytotherapie (1991), 25(1), 32-8.

[159] (a) Alonso, M.L., Montana, F.P., Miranda, M., Castillo, C., Hernandez, J., Benedito, J.L., Interactions between Toxic (As, Cd, Hg and Pb) and Nutritional Essential (Ca, Co, Cr, Cu, Fe, Mn, Mo, Ni, Se, Zn) Elements in the Tissues of Cattle from NW Spain, BioMetals (2004), 17(4), 389-397. (b)

Nishimura, T. and Robins, R.G., A Re-Evaluation of the Solubility and Stability Regions of Calcium Arsenites and Calcium Arsenates in Aqueous Solution at 25 °C, Mineral Processing and Extractive Metallurgy Review (1998),18(3-4), 283-308.

[160] Glew, R.H., Vanderjagt, D.J., Lockett, C., Grivetti, L.E., Smith, G.C., Pastuszyn, A. and Millson, M., Amino Acid, Fatty Acid, and Mineral Composition of 24 Indigenous Plants of Burkina Faso, Journal of Food Composition and Analysis (1997), 10(3), 205-217.

[161] Holzhueter, G., Narayanan, K. and Gerber, T., Structure of Silica in Equisetum arvense, Analytical and Bioanalytical Chemistry (2003), 376(4), 512-517.

[162] Guerin, T., Astruc, A. and Astruc, M., Speciation of Arsenic and Selenium Compounds by HPLC Hyphenated to Specific Detectors: A Review of The Main Separation Techniques, Talanta (1999), 50(1), 1-24.

[163] Takahashi, Y., Ohtaku, N., Mitsunobu, S., Yuita, K., Nomura, M., Determination of the As(III)/As(V) Ratio in Soil by X-ray Absorption Near-Edge Structure (XANES) and its Application to the Arsenic Distribution between Soil and Water, Analytical Sciences (2003), 19(6), 891-896.

[164] Raab, A., Meharg, A.A.; Jaspars, M., Genney, D.R. and Feldmann, J., Arsenic-Glutathione Complexes-Their Stability in Solution and During Separation by Different HPLC Modes, Journal of Analytical Atomic Spectrometry (2004), 19(1), 183-190.

[165] Munson, R.D., Principles of Plant Analysis in Handbook of Reference Methods for Plant Analysis, Ed., Yash P. Kalra, CRC Press LLC, 1998.

[166] Punz, W.F. and Sieghardt, H., the Response of Roots of Herbaceous Plant Species to Heavy Metals, Env. And Exptl. Botany, 1993, 33, 85-98]

[167] Glass, H.J. and Camm, G.S., Characterization of The Correlation between Heavy Metals in The Environment, Journal de Physique IV: Proceedings (2003), 107(XII-th International Conference on Heavy Metals in the Environment, 2003, Volume 1), 549-552.

[168] Markert, B., Interelement Correlations Detectable in Plant Samples Based on Data from Reference Materials and Highly Accurate Research Samples, Fresenius' Journal of Analytical Chemistry (1993), 345(2-4), 318-22.

[169] Montes-Bayon, M., Jeija, J., LeDuc, D.L., Terry, N., Caruso, J.A. and Sanz-Medel, A., HPLC-ICP-MS and ESI-Q-TOF Analysis of Biomolecules Induced in Brassica Juncea During Arsenic Accumulation, J. Anal. At. Spectrom., 2004, 19, 153-158.

APPENDIX: SEQUENTIAL METHOD FOR EXTRACTION AND SPECIATION

OF ARSENIC IN PLANTS

A.1. Plant Sample Collection and Preparation

- Collect Plant samples in plastic bags free from dirt and soil.
- Store collected plant samples in cold ($\geq 4°C$) before washing and drying.
- Wash plant samples thoroughly with copious amounts of distilled water. A final rinse may be given with distilled deionized water (DDW). In case where distilled water is in short supply, samples may be washed thoroughly with clean tap water and finally rinsed with distilled water. Set aside to drain of water for a few minutes and then place on table with clearly marked tag numbers.
- Spread plant samples on table as much as possible to dry quickly and evenly. After thorough air drying, a portion (about 50 g or a suitable amount) of the air-dried plant may be further dried in the oven at about 60-70°C to remove more water and make crisp for grinding. Oven dried samples should be kept air sealed (covering the sample beaker with Parafilm® M, for example) until ready for grinding.
- Plant samples may be ground in a dedicated coffee grinder. Sample should be ground as finely and evenly as possible following the grinding instruction of the coffee grinder. Leafy parts of plants are easy to grind, but hard to grind plant tissues may be soften using mortar and pestle prior to grinding.
- Clean grinder thoroughly between samples. The lid of grinder (usually plastic made) may be washed and wiped with clean paper towel and set aside for drying while taking preparation for the next sample. The grinder cup should be thoroughly dusted with a brush and then wiped with lightly moist paper towel. The worker should avoid breathing the dusted plant particles.
- The ground plant sample may be transferred from the grinder first on a clean piece of paper to aid in putting the powdered plant material in pre-cleaned dry containers. The ground plant samples should be stored in the freezer. Frozen sample should be brought to room temperature before weighing.

A.2. Extraction Procedure

- Accurately weigh 0.25 g ground plant samples into clean and dry 15 mL extraction tubes. The extraction tubes should be numbered to keep track of samples.

A.2.1. Water-methanol extraction

- Introduce 10 mL 1:1 v/v water-methanol (extractant) and close lid securely. Mix solvent and sample thoroughly with a vortex apparatus for one minute. Sonicate the mixture for 20 minutes in the first step of the three step extractions. After sonication, centrifuge at 3,000 rpm for 10 minutes. Decant supernatant solution into a sample vial that will hold at least 30 mL extract. Take care not to transfer plant materials with the extract to the vial. Add again 10 mL extractant to the residue in the extraction tube. Disperse residue into solvent by vortexing, and tapping if needed. Sonicate for 10 minutes in the second step. Centrifuge and decant as in the first step. Repeat the second step for the third extraction of the sample. Some plant materials may float on top of the liquid column; therefore, care should be taken not to transfer it to the sample extract vial.

A.2.2. Sequential Extraction with 0.1 M-HCl

- Repeat the water-methanol extraction procedure (2.1) for the residue using 0.1 M-HCl extractant. Store extracts in separate sample vials. Acid solvents were found to have less wetting property than the alcoholic solvents. This makes some plant materials to float on top of the liquid column; therefore, more care should be taken during decantation to keep plant materials from transferring from the extraction tube to the sample extract vial.

A.2.3. Extraction of Inorganic Arsenic from Plants using 0.1 M-HCl

- Follow the water-methanol extraction procedure using 0.1 M-HCl as extractant. This is three-step single solvent extraction and gives a quicker assessment of the inorganic arsenic.

A.3. Preparation of Extracts for Arsenic Determination

The HCl extracts after filtration can be analyzed directly by HG-AAS and ICP-AES for total arsenic. The arsenic speciation of the HCl extracts can be conducted by HPLC-HGAAS and other methods. However, water-methanol extracts require further preparation before analysis and are described below.

- Remove solvent from the 1:1 water-methanol extracts by placing the vials with open lids in an oven at 60-65°C.

- Regenerate dried extract into solution using appropriate volume of DDW. For the speciation of the trace organoarsenic species by HPLC-HGAAS, small volume (3-5 mL) of DDW may be suitable for preconcentration of the species.

- Filter extracts before ICP and HGAAS determinations. For HPLC-HGAAS determination, however, initial filtration may be skipped since all samples are introduced by 1 mL syringe and syringe filters to into the HPLC injection port.

- The total arsenic in the extracts may be determined by analyzing the extracts directly by ICP-AES or HG-AAS. A preliminary assessment by the ICP-AES method can be used as a guideline for the determination of the low-level arsenic by HG-AAS method.

- Total As in the extracts can also be obtained by adding the concentration of all arsenic species determined by HPLC-HGAAS or HPLC-ICPMS.

For the determination of total arsenic in plant samples, follow the standard procedure as practiced and/or suitable in the particular laboratory setup.

www.ingramcontent.com/pod-product-compliance
Lightning Source LLC
Chambersburg PA
CBHW021814170526
45157CB00007B/2585